PRE-HSE

Core Skills in Mathematics

D1368518

New Readers Press®
ProLiteracy's publishing division

Photos courtesy of:

p. 8: © FMStox; p. 9: © XAOC; p. 60: © Dancestrokes; p. 114: © PRILL; p. 120: Olivier Le Queinec; p. 129: © ericlefrancais; p. 132: © Sebastien Burel; p. 144: © donghero; p. 150: © chromatos

Pre-HSE Core Skills in Mathematics
ISBN 978-1-56420-881-1

Copyright © 2016 New Readers Press
New Readers Press
ProLiteracy's Publishing Division
104 Marcellus Street, Syracuse, New York 13204
www.newreaderspress.com

Printed in the United States of America
10   9   8   7   6   5   4   3   2

Proceeds from the sale of New Readers Press materials support professional development, training, and technical assistance programs of ProLiteracy that benefit local literacy programs in the U.S. and around the globe.

Developer: QuaraCORE
Editorial Director: Terrie Lipke
Cover Design: Carolyn Wallace
Technology Specialist: Maryellen Casey

CONTENTS

CONTENTS

Welcome to *Core Skills in Mathematics,* an important resource in helping you build a solid foundation of math skills as you gear up to start preparing for the GED®, TASC, or HiSET® high school equivalency Mathematics test.

How to Use This Book

Pretest

The first step in using *Core Skills in Mathematics* is to take the Pretest, which begins on the next page. This test will help you identify your mathematics areas of strength and those where you will need more practice. After taking the Pretest and checking your answers, use the chart on page 11 to find parts of the book that will help you focus on the mathematics skills where you need the most practice.

Mathematics Skills Lessons

The book is organized into seven units, each containing brief lessons that focus on specific skills. Each lesson includes sample problems demonstrating the lesson skills. Read each of the lessons, which also include tips to help you understand math skills and concepts:

- Math Facts point out interesting or important information about math concepts.

- Real-World Connections describe how math concepts connect to our everyday lives.

- Skills Tips offer clues for easy ways to remember skills discussed in the lessons.

- Vocabulary Tips provide advice on using and remembering key math terms.

Each lesson is followed by a brief Lesson Review with questions to test your understanding of the lesson content. Answers can be found in the Answer Key, beginning on page 154.

Each unit concludes with a Unit Practice Test that covers all the content in the unit's lessons. In the Answer Key, you will find explanations to help your understanding with many Unit Test questions.

Posttest

After completing all of the lessons, you can practice what you know by taking the Posttest, beginning on page 140. This test will help you check your understanding of all the skills in the book. It will also help you to see if you are ready to move on to high school equivalency test preparation.

Answer the following questions to test your knowledge of math content and skills.

1. Which of the following numbers is the standard form of the number shown here in expanded form?

   $(7 \times 100{,}000) + (4 \times 1{,}000) + (9 \times 1) + (6 \times \frac{1}{100})$

   A. 704,009.06

   B. 704,900.6

   C. 740,009.6

   D. 7,004,009.06

Match each fraction in column A with the equivalent decimal in column B.

Column A	Column B
2. $\frac{4}{5}$	2.8
3. $\frac{14}{5}$	1.8
4. $1\frac{4}{5}$	0.8

Answer the question based on the illustration.

5. What is the time on the clock shown?

   _____

Solve the problem.

6. Gigi bought four pairs of fuzzy socks at the novelty store. She gave the cashier a $20 bill and received $4.60 in change. If each pair of socks cost the same amount, how much did one pair cost?

   A. $2.75

   B. $3.85

   C. $4.60

   D. $15.40

7. Write "Negative five times the difference of y squared and two" as an expression.

   _____

8. Solve the equation $7x - x + 9 = 3x + 18 + x - 20$.

   _____

Fill in the blanks using the diagram.

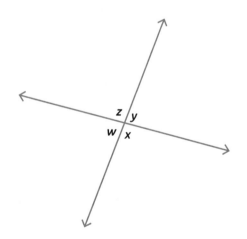

9. If m∠w = 83°, then m∠y = _____

10. If m∠w = 79°, then m∠z = _____

11. Solve the inequality $\frac{y}{11} - 8 > -13$. _____

Answer the questions based on the line graph shown.

The percentage of seats filled by fans for Team Z's games is shown on the graph.

Attendance at Team Z's Baseball Games

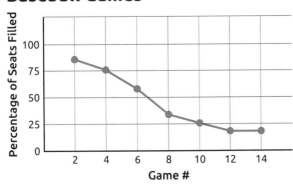

12. During which game were 75% of the seats filled by fans? _____

13. What was the trend in attendance at Team Z's baseball games?

    A. Attendance rose sharply from Game 2 to Game 8, then began to level off.

    B. Attendance fell sharply from Game 2 to Game 8, then began to level off.

    C. Attendance was about equal for all of the games.

    D. The percentage of seats filled remained above 25% for all games.

Answer the question based on the data in the table.

A history teacher records her students' scores on a unit test. Because one student was absent, the teacher records a score of 0 until that student can take the test.

0								
60	63	65	65	68				
70	70	71	73	74	75	78	78	79
81	82	83	83	83	87			
90	92	96						

14. Based on the distribution of the test scores, which of the following statements best describes the measure of center? (NOTE: You may wish to sketch a histogram for the data.)

    A. The mode is the best measure of center because 83 is the most popular score.

    B. The mean will be lower than the median because of the outlier score of 0.

    C. The mean will be higher than the median because of the outlier score of 96.

    D. The median will be more affected than the mean by the outlier score of 0.

Compare the pair of numbers using the number line. Write < (less than) or > (greater than) in the blank to make the statement true.

15. −1 _____ −6

Solve the problems.

16. The asking price of a house was $178,500. The Wilsons paid $101,900 for the house at the closing. By how much did the owners lower the price of the house?

    A. $74,200

    B. $75,800

    C. $76,600

    D. $77,400

17. Harrison wants to buy a sample of latex paint before deciding on a color for his bedroom. A sample that is $\frac{1}{8}$ of a gallon costs $4.35, while a whole gallon of latex paint costs $24.99. Are these rates proportional? Explain. _____

    _____

    _____

18. Rashad follows a recipe that lists the liquid ingredients: water and oil. He mixes $2\frac{3}{4}$ cups of water and $1\frac{2}{3}$ cups of oil. How many cups of liquid does the recipe call for?

    A. $3\frac{12}{17}$

    B. $3\frac{5}{7}$

    C. $4\frac{5}{12}$

    D. $4\frac{7}{12}$

19. Which of the following statements is true?

    A. All quadrilaterals are parallelograms.

    B. A rhombus is also a square.

    C. A rectangle is also a parallelogram.

    D. All parallelograms contain 4 right angles.

20. The triangular yield sign shown has a perimeter of 108 inches. All of its sides are equal, and its area is about 562 square inches. What is the height of the yield sign in inches? Round your answer to the nearest inch.

    A. 36 in

    B. 31 in

    C. 22 in

    D. 18 in

21. How many cubes, each having a volume of 64 cubic centimeters, could fit into a rectangular solid (box) that measures 12 centimeters long by 8 centimeters wide by 16 centimeters high?

    _____

Answer the question based on the illustration.

22. What is the length of the screwdriver in centimeters? _____ cm

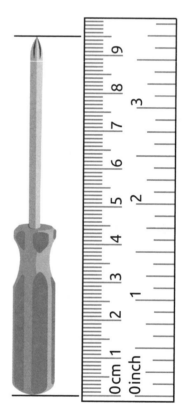

Solve the problem.

23. George is selling his truck for 20% less than the listed value. If he sells the truck for $2,750, what was the listed value?

    A. $2,200.50

    B. $3,437.50

    C. $11,000.00

    D. $13,750.00

Answer the questions based on the bar graph shown.

A high school has four after-school clubs. The number of students in each club is shown on the graph.

Number of Students in Clubs

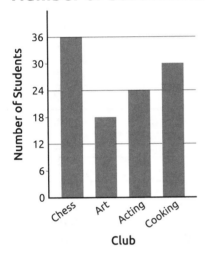

24. How many more students are in Acting Club than in Art Club? _____

25. What percentage of the students in clubs are in Chess Club? Round your answer to the nearest tenth of a percent. _____

26. Convert 2.56 kilometers to meters. _____ m

Solve the problem.

27. Terrence wants to earn at least $350 this summer by mowing lawns. If he earns $25 per lawn, how many lawns must Terrence mow over the summer to reach his goal?

   A. At most 12 lawns

   B. At most 14 lawns

   C. At least 12 lawns

   D. At least 14 lawns

28. Solve the inequality $a + 4.6 \leq 7.2$. _____

29. Evaluate the expression $2.5x + b$ when $x = 4$ and $b = 52.8$. _____

Choose the expression that could be used to solve the problem.

30. Using the food label shown, about how many cups of food are in this container?

Nutrition Facts
Serving Size 1/4 Cup (30g)
Servings Per Container About 38

Amount Per Serving

Calories 200 Calories from Fat 150

	% Daily Value*
Total Fat 17g	**26%**
Saturated Fat 2.5g	**13%**
Trans Fat 0g	
Cholesterol 0mg	**0%**
Sodium 120mg	**5%**
Total Carbohydrate 7g	**2%**
Dietary Fiber 2g	**8%**
Sugars 1g	
Protein 5g	

Vitamin A 0%	•	Vitamin C 0%
Calcium 4%	•	Iron 8%

*Percent Daily Values are based on a 2,000 calorie diet.

A. $38 \div \frac{1}{4}$

B. $\frac{1}{4} \times 38$

C. $\frac{1}{4} + 38$

D. $38 - \frac{1}{4}$

Solve the problem.

31. Vanessa wants to separate three types of candy evenly into goodie bags. She has 35 chocolates, 49 caramels, and 21 lollipops. How many goodie bags can be filled? How many pieces of each type of candy will be in each bag?

   _____ bags

   Chocolates: _____

   Caramels: _____

   Lollipops: _____

Round as indicated.

32. Round 16.8 to the nearest whole number. _____

33. Round 0.942 to the nearest tenth. _____

34. Solve the equation $\frac{N}{19} = 10$. _____

35. Solve the inequality $-5x \geq 10$, and then sketch its graph using the number line provided.

    _____

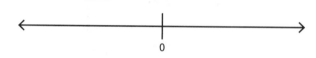

Solve the problem based on the illustration.

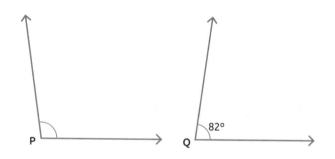

36. If angles P and Q are supplementary angles, then m∠P _____.

Solve the problem.

37. Boris has a beaker with 850 milliliters of saline solution and a beaker with 525 milliliters of distilled water. When he combines the liquids, how many LITERS of liquid will there be in all?

    A. 1,375   **C.** 1.375

    B. 325   **D.** 0.325

38. Convert 31,680 inches to miles.
    Hint: 1 mile = 63,360 inches. _____ mi

Answer these questions based on the data in the table.

The heights, in inches, of eight basketball players in the National Basketball League are listed in the table.

Player	Height (in inches)
Anthony, Carmelo	80
Bryant, Kobe	78
Butler, Jimmy	79
Bynum, Andrew	84
Davis, Anthony	82
Hibbert, Roy	86
Lowry, Kyle	72
Rose, Derrick	75

39. What are the mean, median, and mode of the height data? Round to the nearest tenth, if necessary.

    Mean: _____   Median: _____   Mode: _____

40. What is the range of the data? _____

41. Which of the following statements best describes the measures of center for this data?

    A. The mean is lower than the median because of the outlier 72.

    B. The mean is higher than the median because of the outlier 86.

    C. The mean is the same as the median, so both are useful measures of center.

    D. The mode is the best measure of center for the heights listed.

ANSWER KEY

1. A.
2. 0.8
3. 2.8
4. 1.8
5. 11:55
6. B.
7. $-5(y^2 - 2)$
8. $x = -\frac{11}{2}$ or $-5\frac{1}{2}$ or -5.5
9. 83°
10. 101°
11. $y > -55$ or $-55 < y$
12. Game 4
13. B.
14. B.
15. >
16. C.
17. No, because the cross products of the proportion comparing gallons to cost are not equal. Use 0.125 for $\frac{1}{8}$.

$$\frac{1\,gal}{\$24.99} \overset{?}{=} \frac{0.125\,gal}{\$4.35} \text{ and } \$4.35 \neq \$3.12$$

18. C.
19. C.
20. B.
21. 24
22. 9.5 cm
23. B.
24. 6
25. 33.3%
26. 2,560 m
27. D.
28. $a \leq 2.6$ or $2.6 \geq a$
29. 62.8
30. B.
31. 7 bags; Chocolates: 5; Caramels: 7; Lollipops: 3
32. 17
33. 0.9
34. $N = 190$
35. $x \leq -2$ or $-2 \geq x$

36. 98°
37. C.
38. 0.5 mi or $\frac{1}{2}$ mi
39. Mean: 79.5 inches; Median: 79.5 inches; Mode: There is no mode.
40. 14 inches
41. C.

Check your answers. Review the questions you did not answer correctly. You can use the chart below to locate lessons in this book that will help you learn more about math skills. Which lessons do you need to study? Work through the book, paying close attention to the lessons in which you missed the most questions. At the end of the book, you will have a chance to take another test to see how much your score improves.

Question	Where to Look for Help		
	Unit	Lesson	Pages
1	1	2	15
2, 3, 4	1	6	31
5	6	2	115
6	4	5	76
7	4	1	60
8	4	2	64
9, 10, 11	5	3	94
12, 13	6	4	123
14	7	2	134
15	1	1	12
16	1	4	23
17	2	1	44
18	1	7	35
19	5	1	84
20	5	4	99
21	5	5	103
22	6	1	111
23	2	2	48
24, 25, 26	6	4	123
27	4	5	76
28	4	3	68
29	4	1	60
30	3	1	54
31	1	5	27
32, 33	1	3	19
34	4	2	64
35	4	4	72
36	5	3	94
37, 38	6	3	119
39, 40, 41	7	1	130

Numbers and Operations

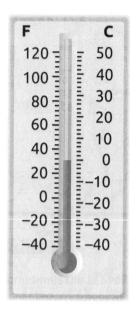

Math is a rich language of numbers and symbols that helps us describe and answer questions about our world. Consider the following situation.

Ai-Ling has a thermometer hanging outside her kitchen window. What is the outdoor temperature in degrees Fahrenheit (F)?

Unit 1 Lesson 1

REPRESENTING NUMBERS ON A NUMBER LINE

Real-World Connection

Negative numbers are often used to show depth. For example, –500 feet describes a submarine that is 500 feet below sea level.

It is important to know the different kinds of numbers we use every day and how these numbers relate to one another.

A **set** is a group of numbers. Some sets of numbers are listed below.

Natural numbers are what we use for counting things: {1, 2, 3, …}. This set is also known as the set of **positive integers**.

Whole numbers are the natural numbers, including the number zero, 0, which represents "nothing." This set is written as {0, 1, 2, 3, …}.

Integers include positive numbers, negative numbers, and zero: {… –3, –2, –1, 0, 1, 2, 3, …}. Note: In a set, the three dots show that the set is infinite, or continues forever.

A **number line** is a tool for viewing sets of numbers. The arrows on the ends of a number line show that the positive and negative numbers continue forever.

A number line helps us compare the size of numbers. A number to the right of another number is greater than that number. A number to the left of another number is less than that number.

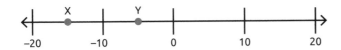

On this number line, point Y is greater than point X because −6 is to the right of −15. We can also say that point X is less than point Y because −15 is to the left of −6.

The symbol < means *is less than*, so we say: −15 < −6 or X < Y

The symbol > means *is greater than*, so we say: −6 > −15 or Y > X

A number line is also useful for finding the distance from one number to another. Tick marks show the spacing between numbers and give each number a location. Look at this example.

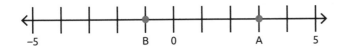

On this number line, notice that there are 5 evenly spaced tick marks between each labeled number. At what number is point A located? At what number is point B located? How far is point B from point A?

ANSWER: Since there are 5 tick marks between 0 and 5, and point A is on the third tick mark to the right of 0, point A is at 3. Similarly, since there are 5 tick marks between 0 and −5, and B is on the first tick mark to the left of 0, point B is at −1.

Find the distance from point B to point A by counting the number of tick marks between them.

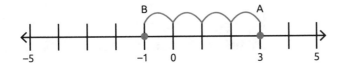

There are 4 tick marks between points A and B, so the distance between A and B is 4 units.

Finally, think again about the thermometer example. Ai-Ling has a thermometer hanging outside her kitchen window. What is the outdoor temperature in degrees Fahrenheit (F)?

Notice that the liquid inside the thermometer lines up with the third short tick mark above 20 on the Fahrenheit side. We need to figure out what the value of this tick mark is to know the outdoor temperature.

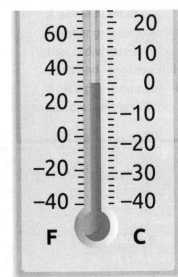

Real-World Connection

32 degrees Fahrenheit is equal to 0 degrees Celsius. On both scales, this is the freezing point of water.

First, focus on the distance between tick marks from 0 to 20 on the thermometer. We can use counting to figure out how many degrees each tick mark represents. Notice that if we count by 4s from 0 to 20, the tick marks would be numbered 0, 4, 8, 12, 16, 20. Similarly, if we count by 4s from 20 to 40, the tick marks would be numbered 20, 24, 28, 32, 40.

So, since the liquid in the thermometer lines up with the third tick mark above 20, it has a value of 32.

ANSWER: The outdoor temperature is 32°F, or 32 degrees Fahrenheit.

Unit 1 Lesson 1

LESSON REVIEW

Complete the activities below to check your understanding of the lesson content. The Unit 1 Answer Key is on page 154.

Skills Practice

Compare numbers using the given number line. Write < (less than) or > (greater than) in each blank to make the statement true.

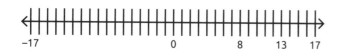

1. −17 ____ 8

2. 0 ____ 13

3. 17 ____ −17

4. 8 ____ 17

Determine the value of each point on the given number line.

5. A = _____

6. B = _____

7. C = _____

Solve the problem.

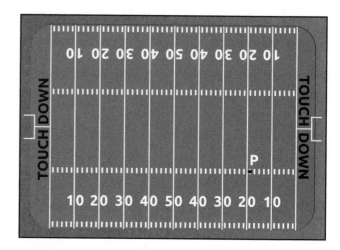

8. A football field measured in yards is shown. If player P crosses the line at the nearest end zone and gets a touchdown, how many yards did he run?

A. 9 yards

B. 10 yards

C. 18 yards

D. 20 yards

UNDERSTANDING PLACE VALUE

We use **place value** in order to understand the structure of our number system. We commonly do math in the **base ten** system using the ten digits 0, 1, 2, 3, 4, 5, 6, 7, 8, and 9 to form multi-digit numbers.

In the **place value chart** that follows, names describe the place, or position, for each digit in a number. Each place to the left of another place in the chart is ten times the other. Each place to the right of another place is one-tenth of the other.

×1,000,000	×100,000	×10,000	×1000	×100	×10	×1	×1/10	×1/100	×1/1000
millions	hundred thousands	ten thousands	thousands	hundreds	tens	ones	tenths	hundredths	thousandths
	,		,			•			

The dot shown to the right of the ones place in the place value chart above is called a **decimal point**. The numbers to the right of the decimal point, called **decimals**, have values less than 1. Decimals have their own place value names, ending in the letters "th."

A ten is the same as a bundle of 10 ones, so we can think of the number 10 as 1 ten and 0 ones. A hundred is a bundle of 10 tens, a thousand is a bundle of 10 hundreds, and so on.

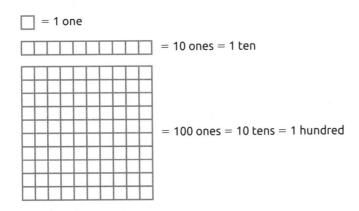

☐ = 1 one

▭▭▭▭▭▭▭▭▭▭ = 10 ones = 1 ten

= 100 ones = 10 tens = 1 hundred

The number 1 is a bundle of 10 tenths. For example, if a candy bar is made up of 10 squares, then a single square is one-tenth of the candy bar.

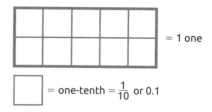

= 1 one

= one-tenth = $\frac{1}{10}$ or 0.1

Math Fact

The commas in the chart separate thousands from hundreds and millions from thousands. Notice that we write a comma after the digit 8 in 8,109.3.

1 is also a bundle of 100 hundredths. For example, a dollar is equal to 100 pennies, so 1 penny is one-hundredth of a dollar.

	× 1,000,000	× 100,000	× 10,000	× 1000	× 100	× 10	× 1	× 1/10	× 1/100	× 1/1000
	millions	hundred thousands	ten thousands	thousands	hundreds	tens	ones	tenths	hundredths	thousandths
				8	1	0	9	3		
	,			,		•				

The **standard form** of the number in the chart above is 8,109.3. Using the place value names for each digit, we can see that the number is made up of 8 thousands, 1 hundred, 0 tens, 9 ones, and 3 tenths.

To show the value of each digit in a number, we can write the number in **expanded form**.

In expanded form, 8,109.3 is written as

$$(8 \times 1{,}000) + (1 \times 100) + (0 \times 10) + (9 \times 1) + (3 \times \tfrac{1}{10})$$

Or, because there are 0 tens, this is simplified as

$$(8 \times 1{,}000) + (1 \times 100) + (9 \times 1) + (3 \times \tfrac{1}{10})$$

Notice that expanded form shows the digit 8 has a value of 8,000, the digit 1 has a value of 100, the digit 9 has a value of 9, and the digit 3 has a value of $\frac{3}{10}$, or 0.3.

Consider the following situation:

While he ran, Quinn's watch recorded his time, 28 minutes and 37 seconds, and the distance he ran, 3.10 miles. Express the number 3.10 in expanded form.

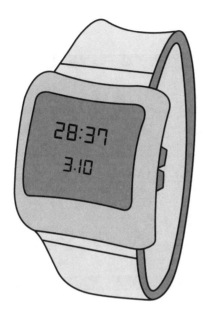

Notice there is a 3 to the left of the decimal point and a 1 to the right of the decimal point. This will help us place the number 3.10 in the chart as follows:

× 1,000,000	× 100,000	× 10,000	× 1000	× 100	× 10	× 1	× 1/10	× 1/100	× 1/1000
millions	hundred thousands	ten thousands	thousands	hundreds	tens	ones	tenths	hundredths	thousandths
						3	1	0	
,			,			•			

ANSWER: 3.10 in expanded form is written as $(3 \times 1) + (1 \times \frac{1}{10}) + (0 \times \frac{1}{100})$, or because there are no hundredths, it can be written more simply as $(3 \times 1) + (1 \times \frac{1}{10})$.

Vocabulary Tip

Decimal comes from the Latin word *decimus,* meaning "tenth." *Decem* means "ten."

Complete the activities below to check your understanding of the lesson content. The Unit 1 Answer Key is on page 154.

Skills Practice

1. Angelica wrote a check to pay her phone bill. Which choice shows the amount of the check in expanded form?

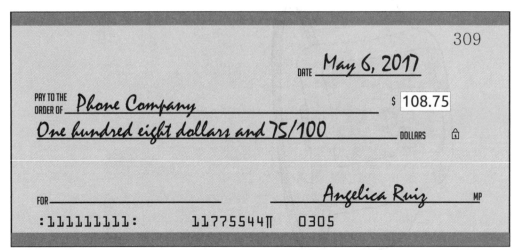

309

DATE _May 6, 2017_

PAY TO THE ORDER OF _Phone Company_ $ 108.75

One hundred eight dollars and 75/100 ____ DOLLARS

FOR ____   _Angelica Ruiz_   MP

:111111111:   11775544⊓   0305

    A. $(1 \times 1{,}000) + (8 \times 100) + (7 \times \frac{1}{10}) + (5 \times \frac{1}{100})$ **C.** $(1 \times 100) + (8 \times 1) + (7 \times \frac{1}{10}) + (5 \times \frac{1}{100})$

    B. $(1 \times 100) + (8 \times 10) + (7 \times \frac{1}{100}) + (5 \times \frac{1}{1000})$ **D.** $(1 \times 100) + (8 \times 10) + (7 \times 1) + (5 \times \frac{1}{10})$

2. Which of the following numbers shows the expanded form $(7 \times 1) + (4 \times \frac{1}{10}) + (1 \times \frac{1}{1000})$ in standard form?

   A. 0.741

   B. 7.401

   C. 7.41

   D. 7,041

4. Write the standard form of the number that has a value of 4 ten thousands, 7 hundreds, 2 tens, 5 ones, and 8 tenths.

   _____

3. The recently discovered jaw of a saber-toothed cat is approximately 5 million years old. Which is the standard form of this number?

   A. 0.0005   **C.** 500,000

   B. 50,000   **D.** 5,000,000

Rounding Numbers

Suppose you buy a new pair of jeans for $19.99. If a friend asks what you paid, you'd probably say "about $20." You have rounded to the nearest ten. Rounding makes numbers easier to work with. We often use rounded numbers in estimates.

Consider this situation:

Adisa takes the California Express train from Chicago, Illinois, to Emeryville, California, traveling a total distance of 2,438 miles. To the <u>nearest thousand</u> miles, about how far does Adisa travel by train?

To answer the question, we can use a place value chart to help us round numbers to a specific place.

First, insert the number you want to round in the place value chart.

millions	hundred thousands	ten thousands	thousands	hundreds	tens	ones	tenths	hundredths	thousandths
			2	4	3	8			

Next, underline the place value you want to round to. To round to the thousands place, underline the 2.

millions	hundred thousands	ten thousands	thousands	hundreds	tens	ones	tenths	hundredths	thousandths
			<u>2</u>	4	3	8			

Look at the digit to the right of the underlined digit.

- If the number is **5 or more**, round up by adding 1 to the underline digit and changing the rest of the digits to the right to zeros.

- If the number is **less than 5**, round down. Keep the underlined digit the same, and change the rest of the digits to the right to zeros.

Vocabulary Tip

To **round** means to change a number to a certain place value so that it is easier to work with.

- Since the number to the right of 2 is 4, round down. Keep the digit 2 the same in the thousands place, and change the rest of the digits to zeros.

- 2,438 rounds to 2,000.

ANSWER: Adisa traveled about 2,000 miles.

Rounding Decimals

Decimals can also be rounded using a very similar method.

Consider this situation:

A zoologist measured a bald eagle's wingspan and found it to be 1.86 meters across. To the <u>nearest tenth</u> of a meter, how wide is the eagle's wingspan?

First, insert the number you want to round in the place value chart.

millions	hundred thousands	ten thousands	thousands	hundreds	tens	ones	tenths	hundredths	thousandths
						1	8	6	

Next, underline the place value you want to round to. To round to tenths, underline the 8.

millions	hundred thousands	ten thousands	thousands	hundreds	tens	ones	tenths	hundredths	thousandths
						1	<u>8</u>	6	

Now, look at the digit to the right of the underlined digit.

- If the number is **5 or more**, round up by adding 1 to the underlined digit and dropping the rest of the digits to the right to zero.

- If the number is **less than 5**, round down. Keep the underlined digit the same and drop the rest of the digits to the right to zero.

Skills Tip

If the place you are rounding up to contains the digit 9, change the 9 to a 0 and add 1 to the place value to its left. Example: <u>9</u>5 rounds up to 100.

- Since the number to the right of 8 is 6, round up. Add 1 to 8 in the tenths place, and drop the rest of the digits to the right to zero.

- 1.86 rounds to 1.9.

ANSWER: The eagle's wingspan is approximately 1.9 meters wide.

Estimating

Rounding is useful when you want to **estimate**, or make an approximate calculation. Suppose you need to stop at the grocery store to buy milk, orange juice, and bread, and you have $10 to spend. The milk costs $3.59, the bread costs $3.78, and the orange juice costs $3.09. Do you have enough money to buy all three items?

Round the price of each item to the nearest dollar. This means to round to the nearest one (or whole number). Underline the ones place in each number, and look at the place to the right (the tenths place) to decide how to round each number.

$\underline{3}.59 \rightarrow$ $3.59 rounds up to $4.

$\underline{3}.78 \rightarrow$ $3.78 rounds up to $4.

$\underline{3}.09 \rightarrow$ $3.09 rounds down to $3.

Now, add the three rounded amounts to get the estimate: $4 + $4 + $3 = $11.

ANSWER: With only $10 to spend, you do not have enough money to buy all three items.

Here's another example of estimation. In the state of Washington, Mount Rainier's elevation is 14,410 feet, while Mount Hood's elevation is 11,239 feet. Estimate how much higher Mount Rainier is compared to Mount Hood.

One way to solve this problem is to start by rounding each elevation to the nearest thousand. Underline the thousands place in each number, and look at the place value to the right (the hundreds place) in order to round.

Mt. Rainier: 1$\underline{4}$,410 rounds down to 14,000 feet.

Mt. Hood: 1$\underline{1}$,239 rounds down to 11,000 feet.

Now, subtract the smaller rounded number from the larger one: 14,000 − 11,000 = 3,000.

ANSWER: Mount Rainier is about 3,000 feet higher than Mount Hood.

Skills Tip

Sometimes it is convenient to round each number in a calculation to the first digit and change the rest of the digits to zeros.

Complete the activities below to check your understanding of the lesson content. The Unit 1 Answer Key is on page 154.

Skills Practice

Round these numbers as directed.

1. 285 to the nearest ten _____

2. 3,443 to the nearest hundred _____

3. 12,148 to the nearest ten thousand _____

4. 1,999 to the nearest thousand _____

5. 84.32 to the nearest tenth _____

6. 0.154 to the nearest hundredth _____

7. 9.5266 to the nearest thousandth _____

8. 3,002.7 to the nearest one (or whole number)

   _____

First, round each dollar amount to the nearest dollar, and then estimate the totals.

9. $8.99 + $4.50 = _____

10. $6.25 + $1.85 = _____

First, round each dollar amount to the nearest ten dollars, and then estimate the totals.

11. $58 + $25 = _____

12. $151 + $19 = _____

Solve the problems.

13. A box contains 775 siding nails. To the nearest ten, how many nails are in the box?

   A. 700

   B. 780

   C. 770

   D. 800

14. A musician sells 1,251,967 copies of her new single. To the nearest ten thousand, how many copies does she sell?

   A. 1,000,000

   B. 1,250,000

   C. 1,252,000

   D. 1,300,000

15. The continent of Australia measures 2,967,892 square miles. The continent of Antarctica measures 5,405,430 square miles. Approximately how many square miles larger is Antarctica than Australia? Use rounding to the nearest million to estimate.

   A. 3,400,000

   B. 3,000,000

   C. 2,400,000

   D. 2,000,000

Operations on Whole Numbers

To combine amounts, we use addition. To find their difference, we use subtraction. As a shortcut for adding the same number many times, we use multiplication. To break a number into equal amounts, we use division.

What is the total shoreline length of the five Great Lakes?

Great Lake	Shoreline Length (in miles)
Lake Superior	2,726
Lake Michigan	1,638
Lake Huron	3,827
Lake Erie	871
Lake Ontario	712

To find the total, add the five lengths together.

- When adding whole numbers, first line up the digits using place value.

$$\begin{array}{r} 2{,}726 \\ 1{,}638 \\ 3{,}827 \\ 871 \\ +712 \end{array}$$

- Line up the ones, tens, hundreds, and thousands so that like place values are added together.

- Start by adding the ones column. Regroup (carry) when necessary.

  For example, when you add all of the digits in the ones column, $6 + 8 + 7 + 1 + 2$, you get 24. Since 24 is equal to 2 tens and 4 ones, carry the 2 to the tens column and write 4 directly below the ones column.

$$\begin{array}{r} {\scriptstyle 3\ 1\ 2} \\ 2{,}726 \\ 1{,}638 \\ 3{,}827 \\ 871 \\ +712 \\ \hline 9{,}774 \end{array}$$

> **Skills Tip**
>
> Addition problems usually contain key words such as *total*, *sum*, or *in all*.

ANSWER: The total shoreline length of the five Great Lakes is 9,774 miles.

PERFORMING OPERATIONS ON WHOLE NUMBERS AND DECIMALS

Operations on Decimals

Mitsuko earned $906.45 last week and spent $32.89 on gasoline for his car. How much of his pay was left after buying gas? To find out how much of his pay was left, subtract.

- When subtracting decimals, first line up the numbers using place value. The decimal points will line up if you do this correctly.

$$\begin{array}{r} 906.45 \\ -32.89 \\ \hline \end{array}$$

- Start by subtracting the ones column. Regroup (borrow) when necessary.

$$\begin{array}{r} {\scriptstyle 3\ 15} \\ 906.4\!\!\!/5 \\ -32.89 \\ \hline 6 \end{array}$$

For example, when you subtract in the ones column, 9 is greater than 5, so borrow 1 ten (10) from the tens column. This changes the 4 in the tens column to a 3.

- Finish subtracting, and be careful when regrouping.

$$\begin{array}{r} {\scriptstyle 8\ 10\ 5\ \ 3\ 15} \\ 9\!\!\!/0\!\!\!/6.4\!\!\!/5 \\ -\ 3\ 2.89 \\ \hline 8\ 7\ 3.56 \end{array}$$

ANSWER: $873.56 of Mitsuko's pay remained after buying gas.

> **Skills Tip**
>
> Use estimation to check your answer: $906.45 is close to $900. $32.89 is close to $30.
> $900 − $30 = $870, so $873.56 makes sense!

Using Multiplication with Decimals

The next example uses multiplication to find the answer.

Jerry can paint an area of 131.25 square feet in one day. If he paints at the same rate, how many square feet can he paint in two weeks (14 days)?

Multiply 131.25 by 14. First, multiply by the ones place (4) in the bottom number. Regroup (carry) when necessary.

$$\begin{array}{r} {\scriptstyle 1\ \ 1\ 2} \\ 131.25 \\ \times\ \ \ \ 14 \\ \hline 525\,00 \end{array}$$

Next, multiply by the digit in the tens place (1) of the bottom number. Since you are multiplying by 10, put a placeholder 0 in the ones column.

$$\begin{array}{r} 131.25 \\ \times\ \ \ \ 14 \\ \hline 52500 \\ 131250 \end{array}$$

Add the partial products to find the answer. Count the number of decimal points in the problem and place that many in your answer.

$$\begin{array}{r} 131.25 \\ \times\ \ \ \ 14 \\ \hline 525\,00 \\ +1312\,50 \\ \hline 1837.50 \end{array}$$

> **Skills Tip**
>
> Be sure to count from the right when placing the decimal point in your answer.

ANSWER: Jerry can paint 1,837.50 or 1,837.5 square feet in two weeks.

Using Division with Decimals

As an example of division with decimals, check the answer to the multiplication problem above.

Jerry painted 1,837.5 square feet in 14 days. To check the answer, divide 1,837.5 by 14 to see if the quotient is 131.25, his painting rate per day.

Set up long division.

There are no decimal places in the divisor, 14, so start doing long division. First, ask yourself, "How many 14s can go into 18?" The answer is 1, so write a 1 above the 8 in the dividend.

Continue dividing as you would with whole numbers. The decimal place in the answer is directly above the location of the decimal point in the dividend.

$$
\begin{array}{r}
131.25 \\
14 \overline{)1837.5} \\
-14 \\
\hline
43 \\
-42 \\
\hline
17 \\
-14 \\
\hline
35 \\
-28 \\
\hline
70
\end{array}
$$

So, the answer checks because the division shows that the rate was indeed 131.25 square feet per day.

Here is another example that requires division of decimals:

Felicia wants to make 1.75-pound boxes of pretzels from a large box of pretzels that weighs 35 pounds. How many boxes of pretzels can she make?

Divide the larger weight, 35 pounds, by the weight of each smaller box, 1.75 pounds.

Set up long division.

Since there are two decimal places in the divisor, place a decimal point to the right of 35 and add two zeros to the right of the decimal point. Now move the decimal point in both numbers two places to the right.

$$
\begin{array}{r}
20 \\
1.75 \overline{)35.00} \\
-350 \\
\hline
0
\end{array}
$$

Divide as you would with whole numbers. The decimal place in the answer is directly above the new location of the decimal point in the dividend.

ANSWER: Felicia can make 20 boxes of pretzels that each weigh 1.75 pounds.

> ## Vocabulary Tip
>
> The number that is being divided is the **dividend**, the number doing the dividing is the **divisor**, and the answer is the **quotient**.

Complete the activities below to check your understanding of the lesson content. The Unit 1 Answer Key is on page 154.

Skills Practice

Solve the problems.

1. Nan wrote checks for $15, $22.75, $189.50, and $75.25. What was the total amount of the checks that she wrote?

    A. $287.50

    B. $287.65

    C. $302.50

    D. $317.65

2. Gasoline prices usually use three digits to show cents. Shown below is the price per gallon and the number of gallons in a sale. Rounding the cost to the nearest cent (two decimal places), how much did the gas cost?

    A. $5.37

    B. $17.20

    C. $39.14

    D. $40.00

3. An auditorium has 1,536 seats. If there are 48 rows of seats, how many seats are in each row?

    A. 22

    B. 28

    C. 30

    D. 32

4. At the beginning of a trip, your car's odometer reads:

| 0 | 3 | 1 | 1 | 7 | 5 |

    At the end of the trip, it reads:

| 0 | 3 | 2 | 4 | 1 | 7 |

    How many miles did you drive?

    _____ miles

FINDING COMMON FACTORS AND MULTIPLES

Multiples

Being able to find common factors and common multiples of two or more numbers is a necessary skill for other important math topics.

Consider this situation:

Starting on the first day of the month, Maria does her laundry every other day, while Ali does his laundry every third day. On which days of the month do both Maria and Ali both do their laundry?

Maria does her laundry on the days in green.

1	2	3	4	5	6	7
8	9	10	11	12	13	14
15	16	17	18	19	20	21
22	23	24	25	26	27	28
29	30					

Ali does his laundry on the days in **bold**.

1	2	3	4	5	6	7
8	9	10	11	12	13	14
15	16	17	18	19	20	21
22	23	24	25	26	27	28
29	30					

Notice that both Maria and Ali do laundry on the 6th, 12th, 18th, 24th, and 30th. These are the dates in common for both Maria and Ali.

A **multiple** of a number is a number that is divided evenly by that number. For example, 50 is a multiple of 10 because $50 \div 10 = 5$, with no remainder.

- All of Maria's numbers are multiples of 2: 2, 4, 6, 8, 10, 12, 14, etc.
- All of Ali's numbers are multiples of 3: 3, 6, 9, 12, 15, 18, etc.

A **common multiple** of two or more numbers is a number that is evenly divisible by all of those numbers.

The laundry dates that Maria and Ali have in common are the common multiples of the numbers 2 and 3, namely 6, 12, 18, 24, and 30. Notice that all of these numbers divide evenly by both 2 and 3.

FINDING COMMON FACTORS AND MULTIPLES

Least Common Multiples

The **least common multiple (LCM)** is the smallest multiple that two or more numbers have in common.

> **BE CAREFUL!**
>
> It is impossible to find the *greatest* common multiple of two or more numbers, since the multiples of numbers continue on forever.

Skills Tip

Sometimes the LCM of two numbers is one of the original numbers. For example, for 4 and 8, the LCM is 8.

The LCM of 2 and 3 is the first number in the list of common multiples: 6. Notice that the LCM of 2 and 3 is their product. This is true sometimes, but not always.

Here is an example of finding the LCM of three numbers. Find the LCM of 5, 10, and 12. List the multiples of each number until you find a common multiple of all three. The first one that is common to all three is the smallest, or least, common multiple.

- Multiples of 5: 5, 10, 15, 20, 25, 30, 35, 40, 45, 50, 55, **60**, 65, 70, …
- Multiples of 10: 10, 20, 30, 40, 50, **60**, 70, …
- Multiples of 12: 12, 24, 36, 48, **60**, …

ANSWER: The LCM of 5, 10, and 12 is 60.

> **BE CAREFUL!**
>
> The LCM of 5 and 10 is 10, but 10 is not the LCM of 5, 10, *and* 12. Remember that you must keep looking at multiples of 12 until you find one in common with the multiples of 5 and 10.

Finding Factors

In contrast, sometimes we need to find the **factors** of a number. These are the numbers that give the original number when multiplied together.

For example, 2 and 7 are factors of 14 because $2 \times 7 = 14$. Also, 1 and 14 are factors of 14 because $1 \times 14 = 14$.

- The factors of 14 are 1, 2, 7, and 14.
- The factors of 21 are 1, 3, 7, and 21.

The **greatest common factor (GCF)** of two or more numbers is the largest factor they have in common.

The GCF of 14 and 21 is 7, because 7 is the largest number in both lists of factors.

To help us find factors of larger numbers, we can use these helpful divisibility rules.

Divisibility Rules	
Divisible by?	**Check!**
2	last digit 0, 2, 4, 6, 8?
3	sum of digits ÷ 3?
4	last 2 digits ÷ 4?
5	last digits 0 or 5?
6	✓2 rule **and** ✓3 rule?
8	last 3 digits ÷ 8?
9	sum of digits ÷ 9?
10	last digits 0?

For example, find the factors of 423. Go through the divisibility rules, starting with the rule for 2.

- 423 does not end in 0, 2, 4, 6, or 8, so it is not divisible by 2.
- $4 + 2 + 3 = 9$, which is divisible by 3, **so 423 is divisible by 3.**
- 23 is not divisible by 4, so 423 is not divisible by 4.
- The last digit, 3, is not a 0 or a 5, so 423 is not divisible by 5.
- 423 is not divisible by 2, so it is not divisible by 6.
- 423 is not divisible by 8.
- $4 + 2 + 3 = 9$, which is divisible by 9, **so 423 is divisible by 9.**
- The last digit, 3, is not a 0, so 423 is not divisible by 10.

Now we can find the factors of 423 by dividing by 9 and by dividing by 3.

$$423 \div 9 = 47$$

$$423 \div 3 = 141$$

FINDING COMMON FACTORS AND MULTIPLES

Continue to find any factors of 47 and 141, which will also be factors of 423. The factors of 423 are: 1, 3, 9, 47, 141, 423.

Here is an application of the GCF. Xiang bought three kinds of candy: 38 caramels, 95 chocolates, and 57 fruit chews. He wants to make goody bags for his daughter's birthday party by dividing each kind of candy equally among the goody bags. How many goody bags can Xiang make? How many of each candy will be in each goody bag?

To solve the problem, divide each type of candy evenly into the same number of goody bags. Start by finding the factors of each number. Use the divisibility rules if you need help.

38: 1, 2, **19**, 38

95: 1, 5, **19**, 95

57: 1, 3, **19**, 57

The GCF of the three numbers is 19, because 19 is the largest factor in common to all three lists.

ANSWER: Xiang can make 19 goody bags. Since 38 ÷ 19 = 2, there will be 2 caramels in each bag. Since 95 ÷ 19 = 5, there will be 5 chocolates in each bag. Lastly, since 57 ÷ 19 = 3, there will be 3 fruit chews in each bag.

LESSON REVIEW

Complete the activities below to check your understanding of the lesson content. The Unit 1 Answer Key is on page 154.

Skills Practice

Solve the problems.

1. Which of the following is the complete list of the multiples of 6?

   A. 1, 2, 3, 6

   B. 6, 12, 18, 24, 30, …

   C. 2, 4, 6, 8, 10, 12, …

   D. 1, 2, 3, 4, 6, 12

2. Find the least common multiple of 9 and 15. Find the greatest common factor of 9 and 15.

   LCM: _____   GCF: _____

3. By which of the numbers 2 through 10 is the number 18,336 divisible?

   A. 2, 3, 4, 6, and 8

   B. 3 and 9

   C. 4, 6, and 8

   D. 2, 3, 4, 8, and 9

4. Fariba visits her sister in the hospital every 6 days. She stops by the flower shop every 16 days. How many days will go by before Fariba goes to the flower shop on the same day she visits her sister in the hospital?

   _____ days

UNDERSTANDING FRACTIONS

Writing Fractions

A **fraction** is a way of representing an amount that is part of a whole. Understanding fractions is important because we often use fractions in our lives.

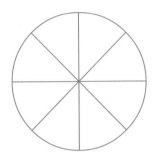

Hans ordered a large vegetable pizza for himself and his three friends. The pizza was cut into 8 equal slices. What fraction of the whole pizza does each person get if each person eats the same amount?

Since the pizza is made up of 8 equal slices and there are 4 people sharing it, each person gets 8 ÷ 4, or 2 slices of pizza. This picture shows the 2 slices that a person will get to eat.

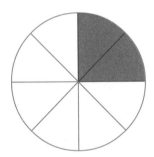

As a fraction, 2 slices out of a total of 8 slices can be written as $\frac{2}{8}$.

The **numerator** of a fraction is the number above the fraction bar. This number represents the part of the whole amount. The **denominator** is the number below the fraction bar. This number represents the whole amount.

In this case, the numerator 2 and the denominator 8 have a common factor of 2. When the numerator and denominator of a fraction share a common denominator, the fraction can be **reduced**, or simplified, by dividing out this common factor. If the fraction is reduced by the GCF, then it is in **lowest terms**.

$$\frac{2 \div 2}{8 \div 2} = \frac{1}{4}$$

So, in simplest fractional terms, each person gets $\frac{1}{4}$ of the pizza.

$\frac{2}{8}$ and $\frac{1}{4}$ are **equivalent fractions** because they have the same value.

Skills Tip

When the numerator and denominator of a fraction are the same but do not equal 0, the fraction is equal to 1. Example: $\frac{3}{3}$ = 1

Using Fractions

Like other numbers, fractions can be represented on a number line. For example, the fraction $\frac{1}{4}$ is one-fourth of the way between the whole numbers 0 and 1.

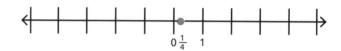

You can also compare fractions using a number line. Notice below that fractions can be negative as well as positive.

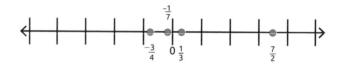

Remember that on a number line, a number to the left of another number is less than (<) that number, and a number to the right of another is greater than (>) that number. The number line shows:

$$-\frac{3}{4} < -\frac{1}{7} \text{ and } \frac{7}{2} > \frac{1}{3}$$

An **improper fraction** is a fraction in which the numerator is larger than the denominator. The fraction $\frac{7}{2}$ is an example. A **mixed number** is a number that has an integer or a whole number part as well as a fractional part. Improper fractions can be rewritten as equivalent mixed numbers, and vice versa.

Converting Fractions

To change $\frac{7}{2}$ to a mixed number, divide 7 by 2 using long division.

The **fraction bar** is a division symbol.

The quotient is 3 remainder 1.

$$\begin{array}{r} 3 \\ 2\overline{)7} \\ -6 \\ \hline 1 \end{array}$$

To form a mixed number, write the whole number of the quotient, 3, and then place the remainder, 1, over the divisor, 2.

$3\frac{1}{2}$

To change $3\frac{1}{2}$ back to an improper fraction, multiply the whole number, 3, by the denominator, 2, and then add the numerator, 1.

$3 \times 2 + 1 = 7$

Write the result, 7, over the denominator, 2.

$\frac{7}{2}$

Math Fact

If the numerator of a fraction is 0 and the denominator is not 0, the fraction is equal to 0. For example, $\frac{0}{35} = 0$.

Finally, it is important to understand that fractions can be converted to decimals, and vice versa. Sometimes it is fairly easy to convert them just by "reading" the number using words.

We read 0.1 as "one-tenth," so the fraction has a numerator of 1 and a denominator of 10: $\frac{1}{10}$.

We read $\frac{9}{100}$ as "nine-hundredths," so the decimal has a 9 in the hundredths place: 0.09.

Other conversions require more work. For example, convert 4.35 to a mixed number.

To change 4.35 to a mixed number, first read the decimal number in words: "four and thirty-five-hundredths." This helps us write the equivalent mixed number.

$$4 \text{ and } 0.35$$

$$4\frac{35}{100}$$

Next, reduce the fraction part to lowest terms. Divide the numerator and denominator by the GCF, 5.

$$4\frac{35}{100} = 4\frac{35 \div 5}{100 \div 5}$$
$$= 4\frac{7}{20}$$

For another example, change $\frac{8}{25}$ to a decimal.

To change $\frac{8}{25}$ to a decimal, divide the numerator, 8, by the denominator, 25, using long division.

Insert a decimal and zeros as needed to the dividend.

$$\begin{array}{r} 0.32 \\ 25\overline{)8.00} \\ -75 \\ \hline 50 \\ -50 \\ \hline 0 \end{array}$$

The quotient is a decimal number less than 1, which makes sense. The fraction has a value less than 1 since its numerator is less than its denominator: 8 < 25.

$$0.32$$

Fractions, mixed numbers, decimals, and integers can all be compared in size on a number line.

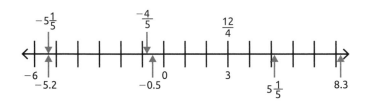

For example, on this number line, $-\frac{4}{5} < -0.5$ and $8.3 > 5\frac{1}{5}$.

Also notice that $-5\frac{1}{5} = -5.2$ and $\frac{12}{4} = 3$.

> **Vocabulary Tip**
>
> A fraction with a denominator of 0 is **undefined**. For example, $\frac{78}{0}$ is undefined.

Complete the activities below to check your understanding of the lesson content. The Unit 1 Answer Key is on page 154.

Skills Practice

Compare numbers using the given number line. Write < (less than), > (greater than), or = (equal to) in each blank to make the statement true.

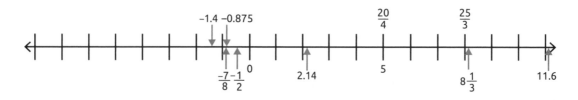

1. $-\frac{1}{2}$ _____ $-\frac{7}{8}$

2. $8\frac{1}{3}$ _____ $\frac{25}{3}$

3. 0 _____ 2.14

4. -1.4 _____ $-\frac{1}{2}$

5. 11.6 _____ -0.875

6. Match each fraction in column A with its equivalent decimal in column B.

A	B
$\frac{1}{100}$	0
$2\frac{4}{5}$	2.22
$\frac{11}{4}$	0.231
$\frac{6}{6}$	2.8
$\frac{0}{19}$	0.01
$\frac{231}{1000}$	2.75
$2\frac{11}{50}$	1

Using Mixed Numbers

Like other numbers, fractions and mixed numbers can be added, subtracted, multiplied, and divided.

A recipe calls for $1\frac{1}{2}$ cups of wheat flour and $\frac{2}{3}$ cup of rice flour. How much flour is combined in this recipe? How much more wheat flour than rice flour is in this recipe? If the recipe is doubled, how much rice flour is needed? If the recipe is cut in half, how much wheat flour is needed?

In this lesson, we will figure out how to answer each of these questions about the recipe.

To answer the first question, we need to add $1\frac{1}{2}$ cups and $\frac{2}{3}$ cup. Notice that the denominators of the fractions in these numbers are *different*.

> **REMEMBER:**
>
> - When adding or subtracting fractions, if the denominators are <u>the same</u>, just add the numerators, and write the sum over the denominator.
>
> - When adding or subtracting fractions, if the denominators are different, rewrite the fractions using the least common denominator, and then add or subtract.

The least common denominator of $\frac{1}{2}$ and $\frac{2}{3}$ is 6 because it is the smallest number evenly divisible by 2 and 3.

$$1\frac{1}{2}+\frac{2}{3}=?$$

Rewrite each fraction with the common denominator of 6. Be sure to multiply each numerator by the same number that you multiplied by in the denominator.

$$1\frac{1}{2}+\frac{2}{3}=1\frac{1\times 3}{2\times 3}+\frac{2\times 2}{3\times 2}$$

$$=1\frac{3}{6}+\frac{4}{6}$$

$$=1+\frac{3}{6}+\frac{4}{6}=1+\frac{3+4}{6}$$

Notice that the whole number in a mixed number can be added separately from the fractions.

$$=1\frac{7}{6}$$

Make sure your final answer is simplified to lowest terms. Change the improper fraction to a mixed number and then add.

$$1\frac{7}{6}=1+1\frac{1}{6}$$

$$=2\frac{1}{6}$$

ANSWER: The recipe combines $2\frac{1}{6}$ cups of flour.

Math Fact

The least common denominator is the same as the least common multiple, LCM, of the denominators.

There are still more questions to answer from the original problem. The second question is, "How much more wheat flour is there than rice flour in this recipe?" To answer this, subtract $\frac{2}{3}$ cup from $1\frac{1}{2}$ cups.

$$1\frac{1}{2} - \frac{2}{3} = ?$$

Again, the denominators are different, so rewrite each fraction with the least common denominator, 6.

$$1\frac{1}{2} - \frac{2}{3} = 1\frac{1 \times 3}{2 \times 3} - \frac{2 \times 2}{3 \times 2}$$

$$= 1\frac{3}{6} - \frac{4}{6}$$

Here it is helpful to change the mixed number to an improper fraction so that the numerators can be subtracted.

$$= \frac{9}{6} - \frac{4}{6}$$

Now the denominators are the same, so subtract the numerators, and write the result over the common denominator.

$$= \frac{9 - 4}{6}$$

$$= \frac{5}{6}$$

ANSWER: There is $\frac{5}{6}$ cup more wheat flour than rice flour in this recipe.

We still have to answer a third question: "If the recipe is doubled, how much rice flour is needed?" To do this, multiply $\frac{2}{3}$ cup by 2.

Vocabulary Tip

When fractions have the same denominators, they are called **like fractions**.

Multiplying fractions is usually easier than adding or subtracting because you don't need to find a common denominator.

$$\frac{2}{3} \times 2 = ?$$

Here it is helpful to change the whole number to a fraction over 1.

$$\frac{2}{3} \times 2 = \frac{2}{3} \times \frac{2}{1}$$

Multiply the numerators. Multiply the denominators.

$$= \frac{2 \times 2}{3 \times 1}$$

$$= \frac{4}{3}$$

Last, change the answer from an improper fraction to a mixed number.

$$= 1\frac{1}{3}$$

ANSWER: Doubling the recipe would mean we need $1\frac{1}{3}$ cups of rice flour.

The last question asks, "If the recipe is cut in half, how much wheat flour is needed?"

To answer the question, divide $1\frac{1}{2}$ cups by 2.

First, rewrite the whole number as a fraction over 1.

$$1\frac{1}{2} \div 2 = 1\frac{1}{2} \div \frac{2}{1}$$
$$= 1\frac{1}{2} \times \frac{1}{2}$$

When dividing fractions, rewrite the problem as a multiplication problem. Change ÷ to × and invert the fraction you are dividing by. In other words, switch the positions of the numerator and the denominator.

Vocabulary Tip

Switching the numerator and denominator of a fraction makes the **reciprocal** of the fraction.

Change the mixed number to an improper fraction.

$$= 1\frac{1}{2} \times \frac{1}{2}$$
$$= \frac{3}{2} \times \frac{1}{2}$$

Multiply the numerators. Multiply the denominators.

$$= \frac{3}{4}$$

The answer is already in simplest form.

ANSWER: Cutting the recipe in half would mean we need $\frac{3}{4}$ cup of wheat flour.

Note that another way of setting this up is to ask the question as, "What is half of $1\frac{1}{2}$?" The key word *of* indicates multiplication, so we would calculate $\frac{1}{2} \times 1\frac{1}{2}$.

Subtracting a Fraction from a Whole

Now let's take a look at an example when a fractional part needs to be subtracted from a whole amount.

Haritha has 30 pounds of clay. She uses $7\frac{3}{4}$ pounds of clay to make a sculpture. How many pounds of clay will she have left after making one sculpture?

Set up the subtraction.

$$30 - 7\frac{3}{4} = ?$$

Write the whole number so that it has the same denominator as the fraction in the mixed number.

$$30 - 7\frac{3}{4} = \frac{30}{1} - 7\frac{3}{4}$$
$$= \frac{30 \times 4}{1 \times 4} - 7\frac{3}{4}$$
$$= \frac{120}{4} - \frac{31}{4}$$

Here it is helpful to change the mixed number to an improper fraction so that the numerators can be subtracted.

Now the denominators are the same, so subtract the numerators, and write the result over the common denominator.

$$= \frac{120 - 31}{4}$$

$$= \frac{89}{4}$$

Change the answer from an improper fraction to a mixed number.

$$= 22\frac{1}{4}$$

ANSWER: $22\frac{1}{4}$ pounds of clay are left after Haritha makes one sculpture.

CHECK: $22\frac{1}{4} + 7\frac{3}{4} = 22 + 7 + \frac{1}{4} + \frac{3}{4} = 29 + \frac{4}{4} = 30$

Complete the activities below to check your understanding of the lesson content. The Unit 1 Answer Key is on page 154.

Skills Practice

Solve the problems.

1. One morning, Penelope walks $\frac{1}{4}$ mile from her home to the zoo. She walks around the zoo and covers $\frac{9}{10}$ mile before walking back home. How many miles does Penelope walk that morning?

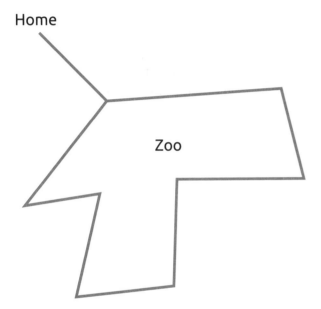

Home

Zoo

   A. $1\frac{3}{20}$

   B. $1\frac{3}{10}$

   C. $1\frac{2}{5}$

   D. $1\frac{3}{4}$

2. A sandwich shop has 48 pounds of sliced turkey. Each 8-inch sub sandwich is made with $\frac{2}{3}$ pound of turkey. How many 8-inch sandwiches can the shop make?

   A. 80

   B. 72

   C. 48

   D. 32

3. During an 8-hour workday, Rashid makes sales calls for $6\frac{2}{5}$ hours. How many hours does Rashid have to do other tasks?

   _____ hours

4. The US Women's National Soccer Team played 24 games in 2014. If the team won $\frac{2}{3}$ of its games, how many games did it win that year?

   _____ games

Answer the questions based on the content covered in this unit. The Unit 1 Answer Key is on page 154.

Solve the problem using the number line.

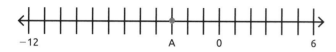

1. What is the value of point A on the number line?

   _____

Compare the numbers using the number line.

2. Write < (less than) or > (greater than) in the blank to make the statement true.

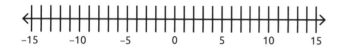

   −9 _____ −14

Solve the problems.

3. Which of the following numbers is the standard form of the number shown in expanded form?

   $(9 \times 10{,}000) + (2 \times 100) + (8 \times 10) + (5 \times \frac{1}{10})$

   A. 92,805

   B. 90,280.5

   C. 9,285

   D. 928.5

4. Write the standard form of the number that has a value of 6 hundred thousands, 4 ten thousands, 2 tens, 9 ones, 3 tenths, and 1 hundredth.

   _____

5. Round 756.24 to the nearest ten.

   _____

6. A stopwatch recorded a precise measurement of 0.931 second. Round this measurement to the nearest hundredth of a second.

   _____ second

7. Helene works part-time. Her time card for last week shows the number of hours she spent at work each day. How many hours did she work in all?

Day	Hours
Monday	5.25
Tuesday	7.75
Wednesday	8
Thursday	7.25
Friday	6.5

_____ hours

8. There will be 351 children at a summer camp. The camp director wants to assign no more than 18 campers per cabin. What is the least number of cabins the camp director needs to have ready for the children?

A. 19

B. 19.5

C. 20

D. 21

9. The original price of a car was $16,895.50. The car dealership discounted the price of the car by $2,540.95. What was the new price of the car after the discount?

A. $14,345.45

B. $14,354.55

C. $14,355.55

D. $19,436.45

10. A teaching staff uses 12.5 reams of copy paper in one week. If each ream has 300 sheets, how many sheets of copy paper did the teaching staff use?

_____ sheets

11. Find the least common multiple (LCM) of 12 and 18. Find the greatest common factor (GCF) of 12 and 18.

A. LCM: 6, GCF: 36

B. LCM: 12, GCF: 9

C. LCM: 9, GCF: 12

D. LCM: 36, GCF: 6

12. Four types of hardware need to be distributed evenly into boxes: 91 screws, 39 nuts, 39 bolts, and 143 nails. How many boxes of hardware can be filled? How many of each type of hardware will fill each box?

_____

_____

_____

_____

Compare numbers using the number line. Write < (less than), > (greater than), or + (equal to) in each blank to make the statement true.

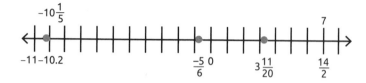

$-10\frac{1}{5}$

7

-11 -10.2

$\frac{-5}{6}$ 0

$3\frac{11}{20}$

$\frac{14}{2}$

13. 7 _____ $\frac{14}{2}$

14. -10.2 _____ 0

15. $-\frac{5}{6}$ _____ $-10\frac{1}{5}$

16. $3\frac{11}{20}$ _____ 7

Choose the equivalent decimal for each fraction.

−1.625   −1.3   −13.125   −0.875

17. $-\frac{13}{10} =$ _____

18. $-13\frac{1}{8} =$ _____

19. $-\frac{7}{8} =$ _____

20. $-\frac{13}{8} =$ _____

21. Three books are stacked in a pile. The first book is $2\frac{1}{2}$ inches tall, the second is $3\frac{7}{8}$ inches tall, and the third is $4\frac{1}{4}$ inches tall. How many inches is the total height of the stack?

   A. $9\frac{7}{8}$

   B. $10\frac{5}{8}$

   C. $11\frac{3}{4}$

   D. $13\frac{1}{4}$

22. A construction worker cuts off the shaded part of a wooden plank shown below. How many feet long is the remaining piece of wood?

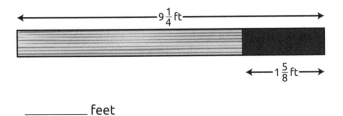

_____ feet

23. Boris owns a car dealership and surveys the vehicles on his lot. Of the 558 vehicles for sale, $\frac{2}{9}$ are trucks. How many vehicles on the lot are NOT trucks?

_____ vehicles

24. A forestry service is dividing 1,206 acres of land into sections that are $\frac{3}{4}$ of an acre each. How many sections can be made from the land?

A. $301\frac{1}{2}$

B. $904\frac{1}{2}$

C. 1,608

D. 1,809

Ratios and Proportional Relationships

Arizona, U.S.

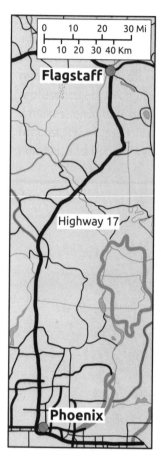

Barry is driving from Flagstaff to Phoenix, and his car's GPS has failed. Fortunately, he brought a backup paper map, so he can still calculate the mileage between stops. Barry approximates that his route is 5 inches long on the map. Using the map scale shown, what is the approximate distance between the two cities, to the nearest mile?

| Write a ratio using the known information in the problem. | Write a proportion with a second ratio using (?) to represent the missing information. | To solve the proportion, multiply the known values that are diagonal from each other and divide by the remaining number. |

$$\frac{1\frac{1}{8}\text{ inches}}{30\text{ miles}}$$

$$\frac{1\frac{1}{8}\text{ inches}}{30\text{ miles}} = \frac{5\text{ inches}}{?\text{ miles}}$$

$$30 \times 5 = 150$$
$$150 \div 1\frac{1}{8}$$
$$= 150 \times \frac{8}{9} \approx 133$$

ANSWER: The distance is about 133 miles.

Unit 2 Lesson 1 — USING RATIOS AND PROPORTIONS TO SOLVE PROBLEMS

A **ratio** is a way to compare two quantities. For example, if there are 41 men and 53 women at a fundraiser, the ratio of men to women can be written as the ratio of 41 to 53, 41:53, or $\frac{41}{53}$.

A **rate** is a special type of ratio that compares two different units. For example, $3.29:5 apples is a rate. A **unit rate** always compares a quantity to 1. For example, 21.5 miles per 1 gallon is a unit rate.

A **proportion** shows that two ratios are equal.

USING RATIOS AND PROPORTIONS TO SOLVE PROBLEMS

Example 1

Determine whether the following ratios are a true proportion.

$$\frac{32}{124} \overset{?}{=} \frac{8}{31}$$

$$32 \times 31 \overset{?}{=} 124 \times 8$$

$$992 = 992$$

Find the diagonal products and see if they are equal.

ANSWER: The diagonal products are equal, so it is a proportion.

> ### Vocabulary Tip
>
> The diagonal products in a proportion are called the **cross products.**

Example 2

Determine whether the following rates are a true proportion.

$$\frac{3.5 \text{ km}}{33 \text{ min}} \overset{?}{=} \frac{5.1 \text{ km}}{52 \text{ min}}$$

$$3.5 \times 52 \overset{?}{=} 33 \times 5.1$$

$$182 \neq 168.3$$

Find the cross products and see if they are equal.

ANSWER: The cross products are not equal, so the ratios are not proportional.

Example 3

Use the rates given below to set up a proportion.

A team of 5 window washers can finish cleaning a skyscraper's windows in 30 hours. If only 3 window washers work on that building, it takes 50 hours.

> ### BE CAREFUL!
>
> This is a bit tricky! It is tempting to write the following proportion:
>
> $$\frac{5 \text{ washers}}{30 \text{ hours}} = \frac{3 \text{ washers}}{50 \text{ hours}}$$
>
> However, the cross products are not equal.
>
> $$5 \times 50 \overset{?}{=} 30 \times 3$$
>
> $$250 \neq 90$$

USING RATIOS AND PROPORTIONS TO SOLVE PROBLEMS

© New Readers Press. All rights reserved.

Skills Tip

Always set up a proportion so that the units in the numerators of both ratios are the same and the units in the denominators of both ratios are the same.

Think of each window washer as doing part of a whole job. When there are 5 washers, each does $\frac{1}{5}$ of the job. When there are 3 washers, each does $\frac{1}{3}$ of the job.

ANSWER: Set up the proportion correctly and check the cross products.

$$\frac{\frac{1}{5} \text{ job}}{30 \text{ hours}} = \frac{\frac{1}{3} \text{ job}}{50 \text{ hours}}$$

Now the cross products are equal.

$$\frac{1}{5} \times 50 \overset{?}{=} 30 \times \frac{1}{3}$$

$$10 = 10$$

Skills Tip

There is more than one correct way to write a proportion. For example:

$\frac{2}{5} = \frac{?}{10}$ and

$\frac{2}{?} = \frac{5}{10}$ and

$\frac{10}{?} = \frac{5}{2}$ are equivalent.

Example 4

Set up a proportion and solve for the missing value.

For a soccer team, the ratio of the team's wins to losses to ties is 11:1:3. If the team continues this pattern, how many games will it win out of 45 games?

The team has won 11 games so far. We need to find out how many games it will win after playing a total of 45 games.

$$\frac{11 \text{ wins}}{? \text{ games}} = \frac{? \text{ wins}}{45 \text{ games}}$$

At first it appears as if there is not enough information to solve this problem. However, we have what we need to find the number of games played so far.

Games played so far equals the number of games won plus the number of games lost plus the number of games tied: $11 + 1 + 3 = 15$.

$$\frac{11 \text{ wins}}{15 \text{ games}} = \frac{? \text{ wins}}{45 \text{ games}}$$

Find the cross product using the two known values. Then, divide by the remaining number.

$$11 \times 45 = 495$$

$$495 \div 15 = 33$$

ANSWER: If the soccer team continues to win at the same rate, it will win 33 out of 45 games.

Complete the activities below to check your understanding of the lesson content. The Unit 2 Answer Key is on page 154.

Skills Practice

Solve the problems.

1. To make a soup, 2 cups of chicken broth are needed for each $\frac{1}{4}$ cup of dried lentils. How many cups of chicken broth are needed if $2\frac{1}{4}$ cups of dried lentils are used?

    _____ cups

2. The scale on a map is 1 inch = 25 miles. What is the actual distance in miles between the water tower and the bridge?

    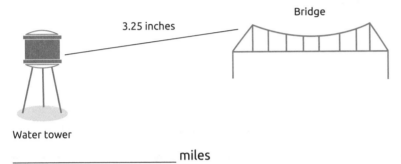

    Bridge

    3.25 inches

    Water tower

    _____ miles

3. The width of the rectangle shown increases from 15 meters to 25 meters. If the length of the rectangle increases proportionally to the width, how many meters is the new length of the rectangle?

    width = 15 m

    length = 35 m

    A. $58\frac{1}{3}$

    B. $55\frac{2}{3}$

    C. 50

    D. 21

4. It takes 8 adults 40 hours to construct a kit house. How many adults would it take to complete a kit house in 32 hours?

    A. 26

    B. 12

    C. 10

    D. 7

SOLVING PERCENT PROBLEMS

Figuring out percents is a part of our daily lives. Consider the following scenario:

Varun recently enjoyed a meal at his favorite café and received the bill shown below. He wants to leave the server a 15% tip. How much should Varun leave for a tip? How much will Varun pay for his meal, including tip?

☘ RESTAURANT

Sale:	$22.87
Tax:	$1.83
Subtotal:	$24.70
Tip:	_____
Total:	_____

An amount such as this tip can be calculated using the **percent equation** below.

$$\textbf{percent} \times \textbf{whole} = \textbf{part}$$

In general, percent problems involve solving for one of these three values.

On Varun's bill, the *whole* is the subtotal, $24.70; the *percent* is 15%; and the *part* is what Varun needs to compute—the tip amount. Varun needs to figure out how much 15% of the subtotal will be. Set up a percent equation and solve for the part.

percent	×	whole	=	part	
15%	×	$24.70	=	part	Substitute known values.
0.15	×	$24.70	=	part	Change percent to a decimal.
		$3.705	=	part	Multiply.
		$3.71	=	part	Round to the nearest cent.

ANSWER: Since the part is actually the tip amount, Varun would leave a $3.71 tip. Notice that the bill from the restaurant has a place to write the tip amount and a place to write the total amount paid for the meal. If Varun adds $3.71 to the subtotal $24.70, his total bill would be $28.41. So, Varun will pay $28.41 for his meal.

In the next example, the percentage and the part are given, and you must calculate the whole.

Janice is scheduled to receive a salary increase of $4.50 per day. This is a $2\frac{1}{4}$ % raise. How much does Janice earn per day before her raise?

The word *increase* is a clue that the amount $4.50 is the part, rather than the whole.

Since you must calculate the whole, use *part ÷ percent = whole* to set up an equation and solve.

part	÷	percent	=	whole	
$4.50	÷	$2\frac{1}{4}$ %	=	whole	Substitute known values.
$4.50	÷	0.0225	=	whole	Change percent to a decimal.
			$200	= whole	Divide.

ANSWER: Therefore, Janice currently makes $200 per day at her job.

Finally, take a look at one more example of a percent problem.

In Game 6 of the 2015 Stanley Cup Final, the Chicago Blackhawks won 42 faceoffs while the Tampa Bay Lightning won 20 faceoffs. What percentage of the faceoffs did the Blackhawks win? Round to the nearest percent.

BE CAREFUL!

It is easy to make the mistake of saying that 42 is the whole and 20 is the part because 42 is greater than 20. However, you must add 42 and 20 to find the total number of faceoffs. The whole = 62. Also, since the problem asks for the percentage of faceoffs won by the Blackhawks, not the Lightning, part = 42, not 20.

The phrase *what percentage* is a clue that the missing value in this example is the percent. Use the equation *part ÷ whole = percent* and solve.

part	÷	whole	=	percent	
42	÷	62	=	percent	Substitute known values.
		0.677	≈	percent	Divide.
		68%	≈	percent	Round to the nearest percent.

ANSWER: The Chicago Blackhawks won about 68% of the faceoffs in Game 6.

Skills Tip

The word *of*, as in "percent of," means the same as *times*, or the multiplication symbol ×.

Math Fact

The percent equation can be also be written as:

part ÷ whole = percent

or

part ÷ percent = whole

SOLVING PERCENT PROBLEMS

Another way to solve this problem is to translate directly from words to math symbols, and then solve the equation.

What percent	of	the faceoffs	was	won by the Blackhawks?
percent	×	whole	=	part
percent	×	62	=	42
		percent	=	42 ÷ 62
		percent	≈	0.68 or 68%

CHECK: Is 68% of 62 equal to 42? Yes, because 0.68 × 62 ≈ 42.

LESSON REVIEW

Complete the activities below to check your understanding of the lesson content. The Unit 2 Answer Key is on page 154.

Skills Practice

Solve the problems.

1. On a quiz with 25 items, there are 8 multiple-choice questions. What percentage of the items are multiple choice?

   A. 3%

   B. 24%

   C. 32%

   D. 68%

2. What is the original price for a gallon of paint?

   _____

Paint
Clearance
75% off
Save $15.00

3. When Juan renewed his cell phone contract, he received a 40% discount off the price of a new phone. He chose a cell phone that regularly sells for $289.50. How much did he pay for the phone?

    A. $115.80

    B. $173.70

    C. $209.60

    D. $249.50

4. The table compares the field goal kicking of four NFL football players in 2014. It shows how many field goals each kicker attempted and how many were made. According to these statistics, which player is the best field goal kicker?

Kicker	Attempts	Makes
C. Barth	16	15
K. Forbath	27	24
S. Gostkowski	37	35
D. Carpenter	38	34

_____

Answer the questions based on the content covered in this unit. The Unit 2 Answer Key is on page 154.

Solve the problems.

1. A summer camp advertises that its ratio of campers to counselors is 7:1. If the camp has hired 37 counselors, what is the maximum number of campers it can host at any one time?

   A. 222

   B. 234

   C. 259

   D. 296

2. The instructions on the back of a package of frozen waffles recommend the following microwave heating time:

   > Heat 5 waffles for 1 min 30 sec

   Shahira wants to heat only 2 waffles. Based on the instructions, for how many seconds should Shahira heat 2 waffles? (Note: 1 min = 60 sec)

   _____ seconds

3. A cyclist on a long-distance trip plans to cover the same distance each day. If he travels at a rate of 18 miles per hour (mph), he will cover the distance shown below in $2\frac{1}{4}$ hours. How many miles is it from point A to point B?

   (Hint: 18 mph can be rewritten as $\frac{18\text{ miles}}{1\text{ hour}}$.)

   A   B

   A. 8

   B. $38\frac{1}{4}$

   C. $40\frac{1}{2}$

   D. 45

4. An 8.2-ounce tube of toothpaste costs $2.99, while a 0.8-ounce travel-size tube costs $0.45. Are these rates proportional? Explain.

   _____

   _____

5. Maya is buying a bathing suit on sale for 35% off the original price. The sale price of the bathing suit is $29.25. What was the original price?

   A. $60.75

   B. $50.00

   C. $45.00

   D. $10.24

6. Gregory found $85\frac{1}{2}$% of the hidden words in a word search puzzle. If there were a total of 200 words in the puzzle, how many words did Gregory find?

   A. 34

   B. 29

   C. 168

   D. 171

7. The table shown gives the season's home run records for three players: Caroline, Mathilda, and Abigail. For each player, find the percent of hits that were home runs. Round to the nearest tenth of a percent, as needed.

Player	Home Runs	Total Hits
Caroline	3	20
Mathilda	5	17
Abigail	8	29

A. Caroline 15%; Mathilda 29.4%; Abigail 27.6%

B. Caroline 6.7%; Mathilda 3.4%; Abigail 3.6%

C. Caroline 23%; Mathilda 22%; Abigail 37%

D. Caroline 19.4%; Mathilda 16.1%; Abigail 26.7%

8. Vukashin ate a meal at a restaurant. He left the server a tip of $4.50. If the tip was 18% of the bill, how much was the bill for the meal?

_____

9. A garden center sells three types of trees—maple, oak, and elm. Currently, there are 130 maple trees, 290 oak trees, and 150 elm trees available. Over Labor Day weekend, the center sells 40% of the maple trees, 10% of the oak trees, and 2% of the elm trees. How many of each tree did the garden center sell over Labor Day weekend?

A. 57 maple trees, 23 oak trees, and 3 elm trees

B. 52 maple trees, 23 oak trees, and 5 elm trees

C. 57 maple trees, 29 oak trees, and 5 elm trees

D. 52 maple trees, 29 oak trees, and 3 elm trees

10. A small coffee shop makes about $1,289 per week. If the shop increases sales by $10 per day, by what percent have sales increased per week? Round your answer to the nearest hundredth of a percent. (Note: 1 week = 7 days.)

A. 5.43%

B. 5.15%

C. 0.78%

D. 0.74%

Operations and Algebraic Thinking

UNIT 3

In 2012, plumbers made an average of $49,140 per year. In 2014, plumbers made an average of $50,660 per year. How much did the average yearly pay increase between 2012 and 2014? If plumbers work 2,080 hours per year, by how much did an average plumber's pay go up per hour?

To answer questions like these, we need to identify the key words or phrases in the problem, translate them to a math equation, and use a step-by-step strategy to solve the equation for the missing information.

Unit 3 Lesson 1 — INTERPRETING AND SOLVING WORD PROBLEMS

Math is a handy tool for problem solving. There are many situations that require us to think logically and approach tasks that might take several steps to finish. It is useful to recognize key words that translate to math and to have a strategy for solving problems.

This chart will help you correctly identify and translate key words to math operations.

Key Word or Phrase	Math Operation
sum, total, plus, altogether, in all, combined, increased by, gained, raised, also, joined	addition +
difference, minus, fewer, take away, less, decreased by, deducted, change in, reduced by, discounted, credit, remain, have left	subtraction −
multiplied by, product of, times, doubled, tripled, twice, of, each, per	multiplication ×
percent, per, quotient of, fraction of, divided by, divided equally, halved	division ÷
is/are/were, the same as, equal to, came to	equal =

Follow this five-step strategy for solving word problems.

Step 1.   Read and understand the problem. Figure out what the question is asking you to find.

Step 2.   Underline or highlight the known facts and information given in the problem.

Step 3.   Choose the correct operations by translating key words to math.

Step 4. Solve the problem. This is usually done by setting up a math **equation** (a statement that quantities have the same value) or by doing the necessary arithmetic. Remember to go back and reread the problem to make sure you answer the original question.

Step 5. Check your answer to see if it makes sense.

Example 1

Mr. Jaffe's gross pay is $800, and his take-home pay is $560. The deductions are what percent of Mr. Jaffe's gross pay?

Step 1. The question asks you to find a percentage. Note: Gross pay is the amount of money earned before some of it is withheld for taxes and other things.

Step 2. Gross pay is $800; take-home pay is $560.

Step 3. The word *deductions* indicates subtraction. The phrase *percentage of* indicates multiplication.

Step 4. Translate the question "The deductions are what percentage of Mr. Jaffe's gross pay?" into a math equation and solve.

First, find the total deductions by subtracting take-home pay from gross pay: $800 − $560 = $240.

Write a percent equation and solve for the percent.

"The deductions are what percent of Mr. Jaffe's gross pay?"

$$\$240 = \text{percent} \times \$800$$

$$\frac{\$240}{\$800} = \text{percent} \qquad \text{Divide by \$800 to solve for the percent.}$$

$$0.3 = \text{percent}$$

$$\text{percent} = 0.3 = 30\% \qquad \text{Convert the decimal to a percent.}$$

ANSWER: The deductions are 30% of Mr. Jaffe's gross pay.

Step 5. Check. Does 30% of $800 = $240?

$$30\% \times \$800 \stackrel{?}{=} \$240$$

$$0.3 \times \$800 \stackrel{?}{=} \$240 \qquad \text{Change the percent to a decimal.}$$

$$\$240 = \$240 \qquad \text{The answer checks.}$$

Skills Tip

Be aware that a problem might contain unnecessary information or not enough information.

Vocabulary Tip

The word *percent* means *out of 100*, so to convert a decimal to a percent, multiply by 100.

INTERPRETING AND SOLVING WORD PROBLEMS

Example 2

April's coffee shop has started selling two kinds of desserts: brownies and cookies. April predicts that she will sell four times as many brownies as cookies. April plans on making a total of 35 desserts for the next morning. Based on her prediction, how many brownies should April bake for the next morning?

Use the five-step strategy to solve this problem.

Step 1.   The question asks how many brownies April should bake.

Step 2.   If April sells 4 times as many brownies as cookies, the number of brownies she sells would equal the number of cookies times 4. For example, if she sells 2 cookies, she will sell 4×2, or 8 brownies.

A total of 35 desserts equals brownies plus cookies.

Step 3.   The key words in the problem are *times*, indicating multiplication, and *total*, indicating addition.

Step 4.   Solve the problem. Set up a math equation using the known information and the correct math operations.

brownies + cookies = 35

In this equation, there are two unknowns. Use the fact that the number of brownies equals four times the number of cookies: brownies = $4 \times$ cookies.

4 $\times$ **cookies** + cookies = 35   Now there is 1 unknown.

(**cookies** + **cookies** + **cookies** + **cookies**) + cookies = 35

5 $\times$ cookies = 35   What number times 5 equals 35?

5 $\times$ 7   = 35   The number of **cookies** = 7.

However, the problem asks how many *brownies* April should bake. Recall that the number of brownies equals 4 times the number of cookies. So, if she bakes 7 cookies, she should bake $4 \times 7 = 28$ brownies.

ANSWER: April should bake 28 brownies.

Step 5.   Check your answer by adding the number of brownies and the number of cookies.

$$\text{brownies} + \text{cookies} \stackrel{?}{=} 35$$

$$28 + 7 \stackrel{?}{=} 35$$

$$35 = 35 \quad \text{The answer checks.}$$

Skills Tip

Remember that multiplication is repeated addition.

For example,
$3 \times 5 = 5 + 5 + 5$ and
$5 \times 3 = 3 + 3 + 3 + 3 + 3$.

Complete the activities below to check your understanding of the lesson content. The Unit 3 Answer Key is on page 155.

Skills Practice

Identify the key word that indicates multiplication.

1. Gigi pays $90 per month for cable TV. She also paid a one-time service fee for installation. What is the total amount Gigi will pay for cable the first year?

   A. per

   B. also

   C. one-time

   D. total

Choose the equation that could be used to solve the problem.

2. A camping group leader had a rope that was 1,185 feet long. She used 82 feet of it to hang food sacks in the trees to keep the food away from bears and 564 feet of it to keep canoes secured. How many feet of rope are left over?

   A. $1,185 - (82 - 564) =$ feet

   B. $1,185 + 82 + 564 =$ feet

   C. $1,185 - (82 + 564) =$ feet

   D. $82 + 564 =$ feet

Solve the problems.

3. Anand sold tickets to a jazz concert. He collected a total of $517.50. The price per ticket was $11.25. How many tickets did Anand sell?

   A. 46

   B. 506

   C. 529

   D. 5,822

4. Pedro had a $20 gift card for a department store. He bought 3 neckties that each cost the same amount. The cost of the 3 neckties plus $4.50 tax after the $20 gift card came to $29.50. How much did each necktie cost?

   A. $45

   B. $18

   C. $15

   D. $1.67

Answer the questions based on the content covered in this unit. The Unit 3 Answer Key is on page 155.

Based on key words, identify the operation needed to solve the problem.

1. A piece of copper pipe is 2.52 meters long. Another pipe is 1.08 meters long. If the pipes are joined, what is the total length?

    A. division

    B. multiplication

    C. addition

    D. subtraction

Choose the equation that could be used to solve the problem.

2. Tabitha owns her own bookstore. If she works 12.25 hours each day, how many hours does she work in a 6-day week?

    A. $12.25 + 6 =$ total hours

    B. $12.25 \times 6 =$ total hours

    C. $12.25 - 6 =$ total hours

    D. $12.25 \div 6 =$ total hours

3. Two of Jupiter's many moons are Ganymede and Europa. It takes Europa 3.551 days to travel around Jupiter, while it takes Ganymede 7.155 days. How many more days does it take Ganymede to travel around Jupiter than it does Europa?

    A. $7.155 \times 3.551 =$ days

    B. $7.155 + 3.551 =$ days

    C. $7.155 \div 3.551 =$ days

    D. $7.155 - 3.551 =$ days

Solve the problems.

4. Juno assembles gift boxes for a cheese shop. Each gift box contains 8 ounces of Swiss cheese, 10 ounces of Havarti cheese, and 9 ounces of cheddar cheese. If Juno assembles 6 gift boxes per hour and he works for 8 hours, how many ounces of cheese will he pack that day?

   A. 1,296

   B. 734

   C. 75

   D. 41

5. In 2014, there were 145,000 organ transplants done in the United States. Of these, 17,105 were kidney transplants. What is the average number of non-kidney transplants done each day? Round your answer to the nearest whole number, if necessary. (Hint: 1 year = 365 days)

   A. 8

   B. 350

   C. 444

   D. 162,470

6. Terrie is 3 times as old as Janelle. The sum of their ages is 24. How many years old is Janelle?

   A. 5

   B. 6

   C. 8

   D. 12

Algebraic Expressions, Equations, and Inequalities

If a skyscraper is 44 stories tall and each story measures x feet high, we can express the height using an algebraic expression.

You have already had some practice solving word problems. In this next unit, we will use algebra to help make the process of solving problems faster and easier. Algebra involves using letters, called variables, to represent missing information. Variables appear in algebraic expressions, equations, and inequalities.

Unit 4 Lesson 1 | EVALUATING EXPRESSIONS

An **algebraic expression** is a combination of math symbols—operators, grouping symbols, numbers, and variables. A **variable** is usually a letter of the alphabet that is used to represent an unknown value.

One important skill to learn in algebra is how to translate words into algebraic expressions.

Take a look at these examples. Notice that variables can be capital or lowercase letters.

Phrase	Operation	Expression
the sum of a number n and 8	addition	$n + 8$
6 plus R		$6 + R$
20 less than B	subtraction	$B - 20$
$3x$ fewer than 15		$15 - 3x$
product of a and b	multiplication	ab
y times 76		$76y$
twice D		$2D$
the opposite of c		$-1 \times c$ or $-c$
n squared		n^2
5 divided by $6m$	division	$5 \div 6m$ or $\frac{5}{6m}$
the quotient of t and 27		$\frac{t}{27}$
4 less than the product of f and g	combination	$fg - 4$
the quotient of p and 7 more than q		$\frac{p}{q+7}$
4,724 increased by n percent of 50		$4{,}724 + n\% (50)$

Here are some examples of how to **evaluate**, or find the value of, algebraic expressions when the values of variables are given.

Example 1

Evaluate the following expressions when $x = 2$ and $y = 4$.

Expression	Substitute Values and Perform Operations	Result
$0.5 + x$	$0.5 + 2$	2.5
$y \div x$	$4 \div 2$	2
$11 + xy$	$11 + 2 \times 4$	19
$-y - x$	$-4 - 2$	-6
$\frac{10x}{y}$	$\frac{10 \times 2}{4}$	5
$y^3 - x^4$	$4 \times 4 \times 4 - (2 \times 2 \times 2 \times 2)$	48

Vocabulary Tip

The **exponential expression** a^b (read "a to the b power") has **base** a and **exponent** b. The exponent tells us that base a should multiply itself b times.

A certain order is followed when performing more than one operation. In the expression $11 + xy$ shown in the chart, when $x = 2$ and $y = 4$, this expression becomes $11 + 2 \times 4$. However, instead of adding 11 and 2 first, multiply 2 and 4 first for a final result of $11 + 8 = 19$.

In mathematics, there is a specific **order of operations**, or rules that describe which operations should be done in what sequence.

Step 1.   Do operations inside grouping symbols: parentheses, brackets, or above and below division bars.

Step 2.   Evaluate all exponents.

Step 3.   Do all multiplication and division from left to right.

Step 4.   Do all addition and subtraction from left to right.

Example 2

What is the value of the following expression when $a = 5$ and $b = 0$?

$$2b + 4(a + b)$$
$$= 2(0) + 4(5 + 0) \quad \text{Substitute values.}$$
$$= 2(0) + 4(5) \quad \text{Add inside parentheses.}$$
$$= 0 + 20 \quad \text{Multiply from left to right.}$$

ANSWER: $= 20$ Add from left to right.

Example 3

Translate the following phrase into an algebraic expression and then find its value when $S = 2$ and $T = -7$.

The quotient of S squared and the sum of T and 11

$$\frac{S^2}{T + 11} \qquad \text{Translate. \textit{Squared} means the exponent is 2.}$$

$$= \frac{2^2}{-7 + 11} \qquad \text{Substitute values.}$$

$$= \frac{2^2}{4} \qquad \text{Do the addition below the division bar.}$$

$$= \frac{4}{4} \qquad \text{Evaluate the exponent.}$$

ANSWER: $= 1$ Divide.

You can use a number line to help you add and subtract when there are negative values.

To add $-7 + 11$, start at 11 and move 7 spaces to the left to arrive at 4.

Also note that $-7 + 11$ can be rewritten as $11 + (-7)$ or as $11 - 7$. All have a value of 4.

BE CAREFUL!

Although you can switch the order of numbers or variables when adding or multiplying, this is not true when subtracting or dividing.

For example, $12 - 4 = 8$, but $4 - 12 = -8$. Also, $12 \div 4 = 3$, but $4 \div 12 = \frac{1}{3}$.

Unit 4 Lesson 1   # LESSON REVIEW

Complete the activities below to check your understanding of the lesson content. The Unit 4 Answer Key is on page 155.

Skills Practice

Match each phrase in column A with its algebraic expression in column B. Write the correct expression in each blank provided.

Column A	Column B
_____ 1. The product of x and 7	$-5J$
_____ 2. The sum of 12 and b	$445(x + y)$
_____ 3. Negative five times J	$g \div h$
_____ 4. The quotient of g and h	$7x$
_____ 5. C to the 3rd power minus 9	$-R - 7S$
_____ 6. Six times the difference of z and 22	$6(z - 22)$
_____ 7. 445 times the sum of $x + y$	$C^3 - 9$
_____ 8. The opposite of R decreased by $7S$	$b + 12$

Evaluate the following expressions for the given values of the variable or variables.

9. $y - 2x$ when $x = 3$ and $y = 9$

   A. -15

   B. -3

   C. 3

   D. 15

10. $29.5 + a^3$ when $a = -1$

    A. 26.5

    B. 28.5

    C. 30.5

    D. 32.5

11. What is the quotient of the sum of p and q and 3 squared when $p = -2$ and $q = 7$?

    A. $-\frac{1}{8}$

    B. $\frac{5}{9}$

    C. 1

    D. 14

SOLVING EQUATIONS

Recall that an equation is a statement that quantities have the same value. For example, the following are all equations.

$$2 + 4 = 6 \qquad \frac{15}{24} = \frac{5}{8} \qquad 9 = 9$$

Sometimes an equation contains a variable that represents an unknown value, and you have to **solve** the equation. In other words, you have to perform operations on the equation in order to find the value of the variable.

To solve an equation, use **inverse operations**, or opposite operations. The goal is to isolate the variable, or get it alone on one side of the equal sign, and get its value alone on the other side.

Equation	Inverse Operations		Solution & Check
$b + 5 = 10$	$b + 5 = 10$ — 5 is being added to b. $b + 5 - 5 = 10 - 5$ — Subtract 5 from both sides. $b = 5$		$b = 5$ $5 + 5 \overset{?}{=} 10$ $10 = 10$ True
$8 = x - 20$	$8 = x - 20$ — 20 is being subtracted from x. $8 + 20 = x - 20 + 20$ — Add 20 to both sides. $28 = x$		$x = 28$ $8 \overset{?}{=} 28 - 20$ $8 = 8$ True
$y(-1) = \dfrac{1}{2}$	$y(-1) = \dfrac{1}{2}$ — y is being multiplied by -1. $y(-1) \div -1 = \dfrac{1}{2} \div -1$ — Divide both sides by -1. $y = -\dfrac{1}{2}$		$y = -\dfrac{1}{2}$ $-\dfrac{1}{2}(-1) \overset{?}{=} \dfrac{1}{2}$ $\dfrac{1}{2} = \dfrac{1}{2}$ True
$3 = \dfrac{T}{0.7}$	$3 = \dfrac{T}{0.7}$ — T is being divided by 0.7. $3 \quad 0.7 = \dfrac{T}{0.7} \quad 0.7$ — Multiply both sides by 0.7. $2.1 = T$		$T = 2.1$ $3 \overset{?}{=} \dfrac{2.1}{0.7}$ $3 = 3$ True

Skills Tip

Think of the equal sign in an equation as the center of a scale. Everything on the left side must balance with everything on the right side.

Some equations require two or more steps to solve them.

1. Combine like terms on each side of the equation.

2. Use inverse operations to get all variable terms on one side.

3. Perform any inverse operations of addition and subtraction.

4. Perform any inverse operations of multiplication and division.

Example 1

Solve the equation $-7 + 2x = 11$.

$$-7 + 2x = 11 \qquad \text{-7 is being added to $2x$.}$$

$$-7 - (-7) + 2x = 11 - (-7) \qquad \text{Subtract -7 from both sides.}$$

$$-7 + 7 + 2x = 11 + 7 \qquad \text{Subtracting a negative is the same as adding a positive.}$$

$$2x = 18 \qquad \text{x is multiplied by 2.}$$

$$\frac{2x}{2} = \frac{18}{2} \qquad \text{Divide both sides by 2.}$$

ANSWER: $x = 9$

As a check, substitute 9 for x in the *original* equation.

$$-7 + 2x = 11$$

$$-7 + 2(9) \overset{?}{=} 11 \qquad \text{Substitute 9 for x.}$$

$$-7 + 18 \overset{?}{=} 11 \qquad \text{Use the order of operations.}$$

$$11 = 11 \qquad \text{True}$$

> ## Skills Tip
>
> When you add, subtract, multiply, or divide a quantity on one side of an equation, you must do the same thing on the other side.

Example 2

Solve the equation $14 - p - p = 6p - 12 + 2$.

$14 - p - p = 6p - 12 + 2$	$-p$ and $-p$ are like terms and -12 and 2 are like terms
$14 + (-p - p) = 6p + (-12 + 2)$	Combine like terms.
$14 - 2p = 6p - 10$	
$14 - 2p + 2p = 6p + 2p - 10$	Add $2p$ to both sides.
$14 = 8p - 10$	Combine like terms $6p + 2p$.
$14 + 10 = 8p - 10 + 10$	Add 10 to both sides.
$24 = 8p$	
$24 \div 8 = 8p \div 8$	Divide both sides by 8.

ANSWER: $3 = p$

As a check, substitute 3 for p in the *original* equation.

$14 - p - p = 6p - 12 + 2$	
$14 - 3 - 3 \stackrel{?}{=} 6(3) - 12 + 2$	Substitute 3 for p.
$14 - 3 - 3 \stackrel{?}{=} 18 - 12 + 2$	Use the order of operations.
$8 = 8$	True

Skills Tip

Subtracting is the same as adding the opposite.

For example, 5 minus $2x$, or $5 - 2x$, is the same as 5 plus negative $2x$, or $5 + (-2x)$.

BE CAREFUL!

Like terms have the same variables raised to the same powers: $5x$, $3.6x$, and $\frac{2}{3}x$ are like terms, <u>but</u> $4x$ and $4y$ are *not* like terms because the variables are different. Constants (numbers) are like terms: 4, 5.9, and $7\frac{1}{10}$ are like terms. Only like terms can becombined by addition or subtraction. Like variable terms are combined by adding or subtracting their **coefficients**—the numbers in front of the variables.

LESSON REVIEW

Complete the activities below to check your understanding of the lesson content. The Unit 4 Answer Key is on page 155.

Skills Practice

Solve each equation. Check the solution by substituting your answer into the original equation.

1. $c - 14 = 22$ $\qquad$ $c = $ _____

2. $\dfrac{d}{3} = -5$ $\qquad$ $d = $ _____

3. $-66 = 6y$ $\qquad$ $y = $ _____

4. $5.9 + t = 7.3$ $\qquad$ $t = $ _____

5. $-7a + 1 = 50$ $\qquad$ $a = $ _____

6. $13 - \dfrac{v}{4} = 15$ $\qquad$ $v = $ _____

7. $2n + 9 = 14 - 3n$ $\qquad$ $n = $ _____

8. $x + 3x + 8 = 18 - x + 30$ $\qquad$ $x = $ _____

An **inequality** is a statement relating expressions using the symbols < (less than), > (greater than), ≤ (less than or equal to), and ≥ (greater than or equal to).

To solve an inequality, use inverse operations as you do when solving an equation. The goal is to isolate the variable, or get it alone on one side of the inequality symbol.

Inequality	Inverse Operations		Solution & Check
$c + 1 < 16$	$c + 1 < 16$	1 is being added to c.	$c < 15$, try 14
	$c + 1 - 1 < 16 - 1$	Subtract 1 from both sides.	$14 + 1 \overset{?}{<} 16$
	$c < 15$		$15 < 16$ True
$4 > x - 2.5$	$4 > x - 2.5$	2.5 is being subtracted from x.	$x < 6.5$, try 6
	$4 + 2.5 > x - 2.5 + 2.5$	Add 2.5 to both sides.	$4 \overset{?}{>} 6 - 2.5$
	$6.5 > x$	Read this as 6.5 is greater than x or x is less than 6.5.	$4 > 3.5$ True
$3y \le -30$	$3y \le -30$	y is being multiplied by 3.	$y \le -10$, try -12
	$\dfrac{3y}{3} \le \dfrac{-30}{3}$	Divide both sides by 3.	$3(-12) \overset{?}{\le} -30$
	$y \le -10$		$-36 \le -30$ True
$\dfrac{n}{9} \ge 7$	$\dfrac{n}{9} \ge 7$	n is being divided by 9.	$n \ge 63$, try 72
	$\dfrac{n}{9} \times 9 \ge 7 \times 9$	Multiply both sides by 9	$\dfrac{72}{9} \overset{?}{\ge} 7$
	$n \ge 63$		$8 \ge 7$ True

Math Fact

Unlike most equations, which have only one value as a solution, the solution of most inequalities is an infinite set of values.

Example 1

Solve the inequality below.

$$20 + \frac{n}{5} - 6 \geq 0$$

$\frac{n}{5} + 14 \geq 0$	Combine like terms.
$\frac{n}{5} + 14 - 14 \geq 0 - 14$	Subtract 14 from both sides.
$\frac{n}{5} \geq -14$	
$\frac{n}{5}(5) \geq -14(5)$	Multiply both sides by 5.

ANSWER: $n \geq -70$

To check whether an inequality is true, you actually can choose from an infinite set of numbers. For example, if $n \geq -70$, choose any number that is greater than or equal to -70 and insert that number into the original inequality. Some numbers you pick will make the math calculations easier than others, so choose wisely.

For this problem, you may want to choose a number for n that's a multiple of 5 $(\ldots -15, -10, 0, 5, 10, 15, 20, \ldots)$ because you will have to calculate $\frac{n}{5}$. Here's how the check step would look if you chose 5 for n.

$20 + \frac{5}{5} - 6 \geq 0$	Substitute 5 for n.
$20 + 1 - 6 \overset{?}{\geq} 0$	Simplify.
$15 \overset{?}{\geq} -70$	True

BE CAREFUL!

When you multiply or divide both sides of an inequality by a negative value, you must *reverse* the direction of the inequality symbol.

$$-R \div -1 > 5.3 \div -1$$
$$R < -5.3$$

Notice how the symbol was reversed from $>$ to $<$.

Math Fact

A negative fraction can be written different ways.

$$-\frac{a}{b} = \frac{-a}{b} = \frac{a}{-b}$$

SOLVING INEQUALITIES

Example 2

Solve the inequality $-4 - 7x \leq -9x + 2$.

$$-4 - 7x \leq -9x + 2$$

$-4 - 7x + 7x \leq -9x + 7x + 2$ Add 7x to both sides.

$-4 - 2 \leq -2x + 2 - 2$ Subtract 2 from both sides.

$$-6 \leq -2x$$

$\dfrac{-6}{-2} \leq \dfrac{-2x}{-2}$ Divide both sides by −2.

$3 \geq x$ Reverse the inequality symbol.

To check whether this inequality is true, choose any number that is less than or equal to 3 and insert that number into the original inequality. Remember that some numbers you pick will make the math calculations easier than others, so choose wisely.

Try 0 since $3 \geq 0$ (and $0 \leq 3$).

$$-4 - 7(0) \overset{?}{\leq} -9(0) + 2 \quad \text{Substitute 0 for } x.$$

$$-4 - 0 \overset{?}{\leq} 0 + 2$$

ANSWER: $-4 \leq 2$ True

Sometimes an equation or an inequality will have no solution. Sometimes an equation or inequality will have a solution that includes all real numbers.

Example 3

$$3p + 7 > 2p + 8 + p$$

$3p + 7 > 3p + 8$ Combine like terms

$3p - 3p + 7 > 3p - 3p + 8$ Subtract 3p from both sides.

ANSWER: $7 > 8$ False

Notice that the final inequality is a false statement. This shows that the original inequality has no solution. However, an answer of $7 < 8$ would show that the inequality's solution includes all real numbers.

Complete the activities below to check your understanding of the lesson content. The Unit 4 Answer Key is on page 155.

Skills Practice

Solve each inequality. Check your solution by substituting an appropriate number into the original inequality.

1. $d - 8 \leq 2$ _____

2. $-\dfrac{z}{14} > 1$ _____

3. $3y \geq -45$ _____

4. $t + 1\frac{5}{9} < 6\frac{2}{9}$ _____

5. $1.5 + 4h < 3.1$ _____

6. $\dfrac{-v}{4} - 6 > 0$ _____

7. $m + 3 - 4m \leq 1 - 4m$ _____

8. $-3x + 10 - 2 \geq 7 - x + 3$ _____

GRAPHING EQUATIONS AND INEQUALITIES

Graphing is a way to visualize the solution of an equation or an inequality.

The graph of the solution of an equation with one variable is usually a single point on a number line.

For example, the solution of equation $2x = -6$, which is $x = -3$, can be graphed as shown below.

However, the graph of the solution of an inequality in one variable has to show that the solution includes more than just one number.

For example, here is the graph of the solution of the inequality $2x > -6$, or $x > -3$.

The graph shows that the solution includes all numbers greater than -3. The endpoint is an open circle, meaning the value, -3, is *not* included in the solution. The arrow points to the right to show that the solutions are infinite—they go on forever.

In general, when graphing the solution of an inequality, the symbols $<$ and $>$ correspond to an open circle on a graph.

The symbols $\leq$ and $\geq$ correspond to a closed circle. A closed circle means that the endpoint *is* included in the solution.

Math Fact

The solution of an equation can be represented as a **solution set**. For example, the solution set of the equation $2x = -6$ is $\{-3\}$.

Math Fact

The solution of an inequality can be represented by **set-builder notation**. For instance, the solution of $2x > -6$ is $\{x \mid x > -3\}$ which is read "x such that x is greater than -3."

Inequality	Solution	Graph
$c + 1 < 6$	$c < 5$	Use an open circle at 5 and an arrow pointing left. A number line from −6 to 6 with an open circle at 5 and an arrow pointing left.
$3y \geq -3$	$y \geq -1$	Use a closed circle at −1 and an arrow pointing right. A number line from −6 to 6 with a closed circle at −1 and an arrow pointing right.
$f - 2 > 2$	$f > 4$	Use an open circle at 4 and an arrow pointing right. A number line from −6 to 6 with an open circle at 4 and an arrow pointing right.
$\dfrac{x}{4} \leq -1$	$x \leq -4$	Use a closed circle at −4 and an arrow pointing left. A number line from −6 to 6 with a closed circle at −4 and an arrow pointing left.

Solve, then graph the equation and inequalities. Show the solution as a set or in set-builder notation.

Example 1

$$7x + 4 = -11$$

$7x + 4 - 4 = -11 - 4$ Subtract 4 from both sides.

$$7x = -15$$

$\dfrac{7x}{7} = \dfrac{-15}{7}$ Divide both sides by 7.

$$x = -2\dfrac{1}{7}$$

ANSWER: The solution set is $\left\{ -2\dfrac{1}{7} \right\}$.

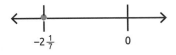

GRAPHING EQUATIONS AND INEQUALITIES

Example 2

$$\frac{a}{5} - 1 < 2$$

$$\frac{a}{5} - 1 + 1 < 2 + 1 \qquad \text{Add 1 to both sides.}$$

$$\frac{a}{5} < 3$$

$$5 \times \frac{a}{5} < 3 \times 5 \qquad \text{Multiply both sides by 5.}$$

$$a < 15$$

ANSWER: In set-builder notation, the solution is $\{a \mid a < 15\}$.

Example 3

$$-9.5y + 4.1y - 6 \geq 5 - y$$

$$-5.4y - 6 \geq 5 - y \qquad \text{Combine like terms.}$$

$$-5.4y + y - 6 \geq 5 - y + y \qquad \text{Add } y \text{ to both sides.}$$

$$-4.4y - 6 \geq 5 \qquad \text{Combine like terms.}$$

$$-4.4y - 6 + 6 \geq 5 + 6 \qquad \text{Add 6 to both sides.}$$

$$-4.4y \geq 11$$

$$\frac{-4.4y}{-4.4} \geq \frac{11}{-4.4} \qquad \text{Divide both sides by } -4.4.$$

$$y \leq -2.5 \qquad \text{Reverse the order of the inequality symbol.}$$

ANSWER: In set-builder notation, the solution is $\{y \mid y \leq -2.5\}$

Make sure you check your solutions!

Math Fact

The **empty set symbol** $\varnothing$ is often used when there is no solution to an equation or an inequality. You can also show an empty number line in this situation.

Complete the activities below to check your understanding of the lesson content. The Unit 4 Answer Key is on page 155.

Skills Practice

Graph the solution of each of the following equations or inequalities on the number line provided.

1. $x \leq 2$

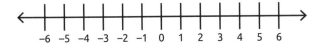

2. $t > -6$

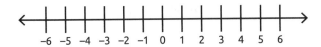

3. $y \leq 5$

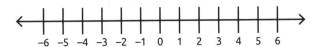

4. $z = 0$

5. $p \geq -3$

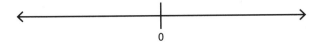

Solve each equation or inequality. Then sketch its graph using the number line provided.

6. $-3q \geq 9$

7. $\dfrac{n}{7} - 1 > 2$

SOLVING PROBLEMS USING EXPRESSIONS, EQUATIONS, AND INEQUALITIES

Skills Tip

When assigning a variable to represent the missing information, it is helpful to choose a letter that reminds you of what it represents. For example, it makes sense to use c for cost, t for time, n for number, and h for height.

Mastering basic algebra skills makes problem solving faster and easier because you can assign a variable to the part that's missing in the problem and solve for that variable to answer the question.

Make sure you still use the five-step strategy for solving word problems introduced earlier.

Step 1. Read and understand the problem. Figure out what the question is asking you to find.

Step 2. Write down the known facts and information given in the problem.

Step 3. Choose the correct operations by translating key words to math.

Step 4. Solve the problem. This is usually done by setting up a math equation or inequality or by doing the necessary arithmetic. Make sure you go back and reread the problem to check that you answered the original question.

Step 5. Check your answer to see if it makes sense.

Example 1

Sammy is saving money to buy a car. So far he has saved $\frac{2}{3}$ of the amount he needs to buy the car. If he has saved $7,800 so far, what is the cost of the car?

Step 1. The question is asking for the cost of the car. Let c = the cost of the car.

Step 2. List the known information.

Sammy has saved $7,800.

$7,800 is $\frac{2}{3}$ of the cost of the car, c.

Step 3. Write an equation.

$7,800 is $\frac{2}{3}$ of the cost of the car

$$\downarrow \quad \downarrow \ \downarrow \ \downarrow \qquad \downarrow$$

$$7{,}800 \ = \frac{2}{3} \times \qquad c$$

Step 4. Solve the equation.

$$7,800 = \frac{2}{3} \times c$$

$$\frac{3}{2} \times 7,800 = \frac{3}{2} \times \frac{2}{3} \times c \quad \text{Multiply both sides by } \frac{3}{2}.$$

$$11,700 = c$$

ANSWER: The cost of the car is $11,700.

Step 5. Since $11,700 is a larger amount than $7,800, this number makes sense for the cost of the car. Check the answer in the original equation.

$$\frac{2}{3} \times 11,700 \overset{?}{=} 7,800$$

$$7,800 = 7,800 \quad \text{True}$$

The answer checks.

Math Fact

For the phrases "at most" and "not more than," use the symbol $\leq$. For the phrases "at least" and "not fewer than," use the symbol $\geq$.

Example 2

Gabrielle is trying to get a B in her biology class. She must earn at least 80% of a possible 2,200 points. If Gabrielle has 1,450 points so far, at least how many additional points must she earn to get a B?

Step 1. Since the question asks "at least," you will need an inequality to find the number of additional points Gabrielle must earn. Let p = the additional number of points.

Step 2. List the known information.

The possible (greatest) number of points a student can earn is 2,200.

Gabrielle has 1,450 points so far and can earn p more points.

80% of 2,200 will earn a B.

ANSWER: 1,450 plus p must be greater than or equal to 80% of 2,200 for a B.

Step 3. Since she needs <u>at least</u> p points, write an inequality.

1,450 points plus p more is at least 80% of 2,200

$$\downarrow \qquad\qquad \downarrow\ \ \downarrow \qquad\quad \downarrow \qquad \downarrow\ \ \downarrow\ \ \ \downarrow$$
$$1{,}450 \qquad + \ \ p \qquad\quad \geq \qquad 0.80\ \ \bullet\ \ 2{,}200$$

Step 4. Solve the inequality.

$$1{,}450 + p \geq 0.80 \bullet 2{,}200$$
$$1{,}450 + p \geq 1{,}760 \qquad\qquad \text{Simplify.}$$
$$1{,}450 - 1{,}450 + p \geq 1{,}760 - 1{,}450 \qquad \text{Subtract 1,450 from both sides.}$$
$$p \geq 310$$

ANSWER: Gabrielle has to earn at least 310 more points to get a B in biology class.

Step 5. Check the answer in the original inequality. Then, check another value that should make the inequality true. For example, $350 \geq 310$.

$$1{,}450 + 310 \overset{?}{\geq} 0.80 \bullet 2{,}200 \qquad\qquad 1{,}450 + 350 \overset{?}{\geq} 0.80 \bullet 2{,}200$$

$$1{,}760 \geq 1{,}760 \quad \text{True} \qquad\qquad\qquad 1{,}800 \geq 1{,}760 \quad \text{True}$$

Both amounts check.

Skills Tip

Remember to change a percent to a decimal or fraction so you can do calculations. For example, 7% equals 0.07 or $\frac{7}{100}$.

Complete the activities below to check your understanding of the lesson content. The Unit 4 Answer Key is on page 155.

Skills Practice

Choose the equation or inequality that can be used to solve the problem.

1. Yann paints two rooms. The first room takes twice as long to paint as the second. Yann finishes both rooms in $7\frac{1}{2}$ hours, including breaks totaling $1\frac{1}{2}$ hours. How many hours does Yann spend painting the second room?

    A. $2h - h = 1.5$

    B. $2h + 1.5 = 7.5$

    C. $h + 2h = 7.5$

    D. $h + 2h + 1.5 = 7.5$

2. Ethan's piano instructor told him to practice at least 20 hours each week to prepare for an upcoming concert. Ethan plans to spend the same amount of time practicing each day all week. Which inequality shows how much time Ethan should practice each day?

    A. $7h \le 20$

    B. $7h \ge 20$

    C. $20h \ge 7$

    D. $20h \le 7$

Solve the problems.

3. Manuela bought 5 identical fan belts at the auto parts store. She gave the cashier a $10 bill and got $3.20 change back. How much did each fan belt cost?

    A. $1.36

    B. $2.15

    C. $6.80

    D. $13.20

4. Rajiv wants to hire a personal trainer at his gym. He has at most $75 to spend on a training session, including a 20% tip based on what the trainer charges. What is the greatest amount that the trainer could charge per session?

    A. $15.00

    B. $62.50

    C. $82.50

    D. $90.00

Answer the questions based on the content covered in this unit. The Unit 4 Answer Key is on page 155.

Match each phrase in column A with its algebraic expression in column B. Write the correct expression in each blank provided.

Column A	Column B
_____ 1. 8 added to y	$h^4 - 6$
_____ 2. h to the 4th power minus 6	$24.5 - 7b$
_____ 3. 24.5 decreased by $7b$	fg^2h
_____ 4. The quotient of x and the sum of $2y$ and z	$y + 8$
_____ 5. The product of f, g squared, and h	$\dfrac{x}{2y+z}$

Solve the problems.

6. Evaluate the expression $a - 0.4b$ when $a = 12$ and $b = 10$.

   A. 1.6

   B. 5.2

   C. 8

   D. 16

7. What is the value of "eight fewer than the product of r and s" when $r = 5$ and $s = 7$?

   A. −27

   B. −16

   C. 10

   D. 27

Solve each equation. Check your solution by substituting your answer into the original equation.

8. $19 + n = 25$ $n =$ _____

9. $1.5x = -3$ $x =$ _____

10. $a - \frac{1}{4} = 9\frac{1}{4}$ $a =$ _____

11. $8b - 1 = 31$ $b =$ _____

12. $\frac{p}{6} - 12 = 17$ $p =$ _____

13. $9 - 5t = -3t - t - 14$ $t =$ _____

14. $1 - w + w = 4 + 18 + 7w$ $w =$ _____

Solve each inequality. Check your solution by substituting an appropriate number into the original inequality.

15. $s + 7.4 > 11.9$ _____

16. $y - 3 \geq 19$ _____

17. $85 - 10b < 95$ _____

18. $-\frac{p}{8} + 1 \leq -1$ _____

19. $-9 + c + 3 > 7 + 5c$ _____

20. $5r - 10 + 2r < -4r + 1$ _____

Solve each equation or inequality. Then, sketch its graph using the number line provided.

21. $7 + a > 9$

22. $x - 14 = 1$

23. $-\dfrac{c}{2} + 5 \geq -3$

24. $-18 < 36y$

Choose the equation or inequality that can be used to solve the problem.

25. One full water cooler bottle holds about 18.9 liters of water. Melinda fills 5 bottles of the same size from the water cooler. After filling her bottles, the water cooler contains 9.8 liters of water. How many liters of water does each of Melinda's bottles hold?

   A. $5w + 9.8 = 18.9$

   B. $18.9 + 5w = 9.8$

   C. $18.9 - 5w = 9.8$

   D. $5w - 18.9 = 9.8$

26. Paulina wants to take a taxi instead of the bus, but she doesn't want to spend more than $5 (not including tip). The taxi costs a flat fee of $3.50 plus $0.25 per mile. How many miles can Paulina ride in the taxi for $5?

   A. $\$3.50 - 0.25m \geq \$5$

   B. $\$3.50 + 0.25m \leq \$5$

   C. $\$5 + 0.25m \geq \$3.50$

   D. $\$5 - 0.25m \leq \$3.50$

Solve the problems.

27. Charlie is thinking of a number. He says that if you divide his number by −4 and subtract 3, the result will be −15. What number is Charlie thinking of?

    A. −63

    B. −48

    C. 48

    D. 63

28. Next semester, 32 college students are going on an overseas trip to do scientific research. The students must contribute at least $22,080 of the $70,016 cost of the trip. The college will pay the remaining cost. If each student contributes an equal share of the cost (c), how much does each student need to pay?

    A. $c \geq \$690$

    B. $c \leq \$690$

    C. $c \geq \$2,188$

    D. $c \leq \$2,188$

UNIT 5

Geometry

In this unit, you will learn to identify geometric shapes and their characteristics. You will also discover the special relationships between angles. Finally, you will solve real-world problems by calculating the perimeters and areas of 2-dimensional figures, as well as the surface areas and volumes of 3-dimensional objects.

Unit 5 Lesson 1

KNOWING SHAPES AND THEIR ATTRIBUTES

In our world, we relate to objects that have the three dimensions of length, width, and height. When we study geometry, we also learn about objects that have zero, one, or two dimensions. Take a look at the following table.

Object	Attributes	Example
Point	A **point** is a location in space, not an actual "thing."	A Point A
Line	A **line** is a straight mark that extends in both directions forever. It has length but no width or thickness. A line can be defined by two points in space.	C D Line CD or $\overleftrightarrow{CD}$
Ray	A **ray** is a line that extends in only one direction forever. It has an endpoint that is either open or closed.	E Ray E or $\overrightarrow{E}$
Line segment	A **line segment** is a line that has two endpoints that can be either open or closed.	F G Segment FG or $\overline{FG}$
Angle	An **angle** is formed by two rays, called the sides of the angle, with a common endpoint, called the **vertex**.	J H I Angle HIJ, angle JIH, angle I, or ∠I (vertex at I)
Curve	A **curve** is a rounded mark with length but no thickness.	K Curve K
Plane	A **plane** is a flat, 2-dimensional surface with no thickness that extends in all directions forever.	L Plane L

KNOWING SHAPES AND THEIR ATTRIBUTES

More About Angles

Angles are often measured in degrees. A **degree** is the amount of counterclockwise rotation of one side of the angle pivoting around its vertex. The degree symbol is °.

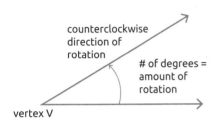

counterclockwise direction of rotation

# of degrees = amount of rotation

vertex V

If one side of an angle rotates all the way around, it rotates 360°.

If it rotates one-fourth of the way around, it rotates $\frac{1}{4}(360°) = 90°$.

If it rotates one-half of the way around, it rotates $\frac{1}{2}(360°) = 180°$.

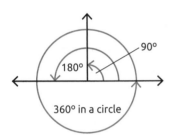

90°

180°

360° in a circle

Vocabulary Tip

A **right angle** measures 90°. A **straight angle** measures 180°. An **acute angle** measures less than 90°. An **obtuse angle** measures greater than 90°. Draw one of each of these angles.

KNOWING SHAPES AND THEIR ATTRIBUTES

Real-World Connection

We can think of train tracks as parallel lines. The rails run alongside each other and are the same distance apart. On the other hand, the corner of a desk is formed by lines that are perpendicular. Look around for other examples.

More About Lines

Two lines in space can create three different situations:

- The lines can cross at exactly one point. This point is called their **intersection**.
- The lines can each extend forever without crossing. Lines that never cross are called **parallel lines**.
- The lines can cross to form a right angle. Lines that intersect and form a right angle are called **perpendicular lines**.

These terms are used a lot when describing 2-dimensional figures.

2-Dimensional Figures

There are many 2-dimensional shapes that have special names. They are defined by their parts, namely, the line segments, or **sides,** that form them and the angles where the sides meet.

Pay attention to the characteristics shared by the following shapes, as well as the ones that set them apart from each other.

Shape	Attributes	Examples
Triangle	<ul><li>It has 3 sides and 3 angles.</li><li>A **right triangle** has one angle that is 90°.</li><li>Angles in a triangle can be right, acute, or obtuse.</li></ul>	Triangle ABC
Quadrilateral	<ul><li>It has 4 sides and 4 angles.</li><li>Sides are *not necessarily* equal.</li><li>Angles are *not necessarily* equal.</li><li>Angles are *not necessarily* right angles.</li></ul>	Quadrilateral DEFG
Rectangle	<ul><li>Opposite sides have equal lengths.</li><li>All 4 angles are right angles. (The small marks on the diagram show the equal sides and right angles.)</li><li>Opposite sides are parallel.</li></ul>	Rectangle HIJK

Rhombus	• All 4 sides have equal length. • Opposite angles are equal. • Opposite sides are parallel.	Rhombus LMNO
Square	• All 4 sides have equal length. • All 4 angles are right angles. • Opposite sides are parallel.	Square PQRS
Parallelogram	• It has 4 sides and 4 angles. • Opposite sides are parallel. • Opposite sides are equal. • Opposite angles are equal.	Parallelogram TUVW
Trapezoid	• It has 4 sides. • It has 1 pair of opposite sides that are parallel.	Trapezoid WXYZ
Pentagon	• It has 5 sides. • If it is a **regular pentagon**, then all sides are equal, and all angles are equal and measure 108°.	
Hexagon	• It has 6 sides. • If it is a **regular hexagon**, then all sides are equal, and all angles are equal and measure 120°.	
Circle	• It is a curved figure (no straight sides) formed by a 360° angle. • It has a center point. • The **diameter** is a segment that has 2 endpoints on the circle and passes through the center of the circle. • The **radius** is a segment that has 1 endpoint at the center of the circle and one on the circle.	Circle T

Vocabulary Tip

In a closed figure with straight sides, the point where two sides form an angle is called a **vertex**. The plural form of vertex is **vertices**.

Complete the activities below to check your understanding of the lesson content. The Unit 5 Answer Key is on page 156.

Skills Practice

Answer the questions.

1. Which of the following statements is true?

   A. A line segment can be measured, but a ray cannot be measured.

   B. Two parallel lines cross at right angles.

   C. The radius of a circle is twice as long as the diameter of the circle.

   D. A plane has three dimensions.

2. Which of the following statements is true?

   A. A triangle is an example of a trapezoid.

   B. A square is also a quadrilateral.

   C. A rhombus is an example of a square.

   D. All quadrilaterals are parallelograms.

3. Which of the following statements is true?

   A. A rhombus is also a hexagon.

   B. A rhombus is also a rectangle.

   C. A rhombus is also a square.

   D. A rhombus is also a parallelogram.

4. Which geometric shapes have all right (90°) angles?

   A. rhombuses and squares

   B. triangles and parallelograms

   C. squares and rectangles

   D. rectangles and trapezoids

In the lesson on shapes, we learned that a plane is a flat surface that extends in all directions forever. A **coordinate plane** is a plane specially designed to show the locations of points, lines, and other geometric objects.

A coordinate plane, or **coordinate grid**, is used to identify the location of a point (x, y) defined by its distance from the intersection of the **x-axis** and the **y-axis**, called the **origin**. The origin is located at point $(0, 0)$.

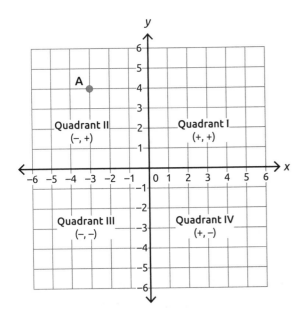

The point (x, y) is in the form of an **ordered pair**. The first number, or **x-coordinate**, tells us how far right or left to travel along the x-axis, starting from the origin. The second number, or **y-coordinate**, tells us how far to move up or down along the y-axis.

For example, point A is located at coordinates $(-3, 4)$ because it is 3 units to the left and 4 units up from the origin.

The plane is divided into 4 areas, called **quadrants**, based on the signs of the coordinates. For example, point A (shown above) is in Quadrant II. Its x-coordinate is negative, and its y-coordinate is positive.

USING A COORDINATE PLANE

Example 1

Plot point B (0, −5) on a coordinate plane.

Start with your pencil at the origin, (0, 0). This point is located where the *x*- and *y*-axes cross. Because the *x*-coordinate of the point is 0, do not move either to the right or left.

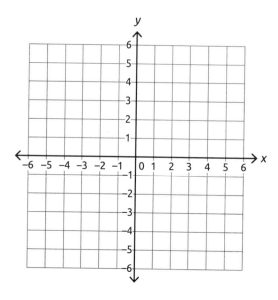

Now, because the *y*-coordinate is −5, move your pencil 5 units down and draw a dot and label it B. Your point will lie on the *y*-axis. Notice that it is not in any specific quadrant.

Example 2

Segment PQ is graphed on the coordinate grid. What are the coordinates of the endpoints? In which quadrant does segment PQ lie?

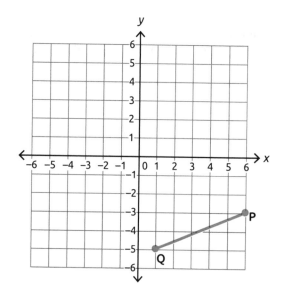

ANSWER: First, determine the coordinates of point P. Starting at the origin, move 6 units to the right, then move down 3 units. The coordinates of P are (6, −3).

Next, find the coordinates of Q. Starting at the origin, move 1 unit to the right and 5 units down. The coordinates of Q are (1, −5).

Segment PQ lies in Quadrant IV.

Example 3

Three vertices and two sides of square HIJK are drawn on the coordinate plane. At what coordinates should point K be located to complete the square?

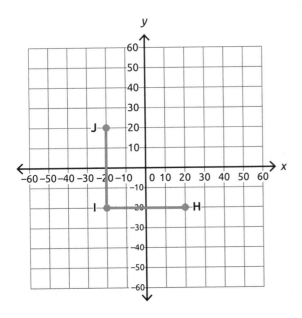

ANSWER: Since this is a square, all four sides need to have the same length. Notice that each tick mark is worth 10 units, so there are 40 units between points H and I and points I and J.

The length of each side of square HIJK is 40 units.

Count 40 units to the right of J to place point K in the correct location. Draw a dot at this point, and then draw a line connecting J and K and another line connecting K and H.

Now, to answer the original question, what are the coordinates of point K? Starting at (0, 0), to get to point K, you would move 20 units to the right and 20 units up. So, the x-coordinate of point K is 20, and the y-coordinate of point K is 20, or (20, 20).

Skills Tip

The x- and y-axes are number lines, so between two tick marks labeled with integers, there are unlabeled decimals and fractions. For example, halfway between 0 and 1 on either axis would be 0.5, or $\frac{1}{2}$.

USING A COORDINATE PLANE

Example 4

This coordinate grid shows the layout of a city's streets. Belinda's apartment building is at point B.

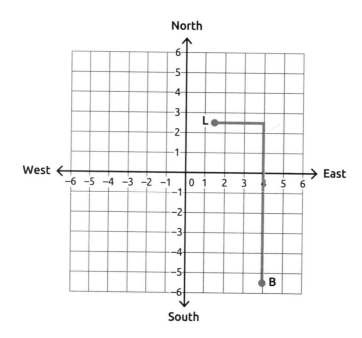

Belinda walks 8 blocks due north, then 2.5 blocks due west to the library. What are the coordinates of the library on the map?

ANSWER: Starting at B, if Belinda walks 8 blocks due north, and then 2.5 blocks due west, she arrives at point L, the location of the library.

To find the coordinates of point L, start at the origin (0, 0) and move 1.5 units (blocks) to the right, then 2.5 units up. L has coordinates (1.5, 2.5).

Complete the activities below to check your understanding of the lesson content. The Unit 5 Answer Key is on page 156.

Skills Practice

Give the coordinates of the indicated points.

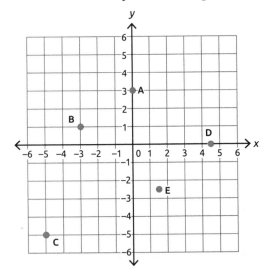

1. Point A (_____ , _____)

2. Point B (_____ , _____)

3. Point C (_____ , _____)

4. Point D (_____ , _____)

5. Point E (_____ , _____)

Plot the points in questions 6–10 on the coordinate plane.

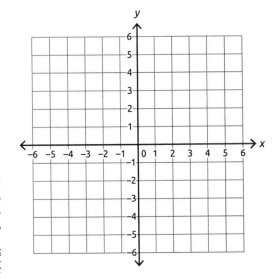

6. Point F at (−4, 1)

7. Point G at (0, 6)

8. Point H at (−2, 0)

9. Point I at (3.5, −3)

10. Point J at (−1.5, −4.5)

Solve the problems.

11. On a coordinate plane, point Y is located at (−800, 100). If point Z is 20 units below and 975 units to the right of point Y, what are the coordinates of point Z?

_____

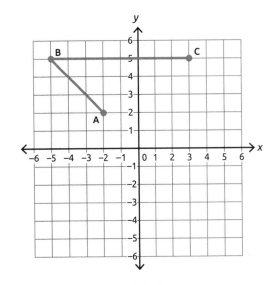

12. Three vertices and two sides of parallelogram ABCD are drawn on the coordinate grid. At what coordinates should point D be located?

A. (3, 2)   C. (6, 2)

B. (3, 5)   D. (6, 5)

Lesson 2 | Using a Coordinate Plane

SOLVING ANGLE MEASURE PROBLEMS

Recall that an angle is made up of two rays with a common endpoint called a vertex. The angle on the right can be named angle HIJ, angle JIH, angle I using its vertex, or angle *i* using the small letter between the two sides. You can also use a symbol to name the angle: ∠HIJ, ∠JIH, ∠I, or ∠*i*.

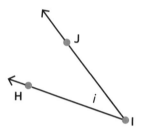

Remember that an angle can be measured in degrees; the symbol for degrees is °. If one side of an angle rotates around its vertex in a full circle, it has rotated 360°. Review the following information about specific angle measures.

Angle Name	Angle Measure	Example
Straight angle	180°	
Right angle	90°	
Acute angle	Less than 90°	
Obtuse angle	Greater than 90°	

In this lesson, you will learn how two or more angles are related based on their location and size.

Take a look at these two intersecting lines.

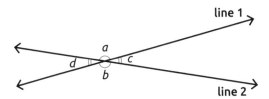

The pairs of angles across from each other formed by these lines are called **vertical angles** (*a* and *b* form one pair; *c* and *d* another pair).

An important **theorem**, or proven principle, follows.

The symbol for congruence is ≅: ∠*b* ≅ ∠*a* and ∠*c* ≅ ∠*d*.

So, if you know that the measure of angle *a* is 155°, or m∠*a* = 155°, then you know that m∠*b* is also 155°. And, if you know that m∠*c* is 25°, then you know that m∠*d* is 25°. Note that the matching marks on the diagram show the congruence of the pairs of angles.

Using the same diagram, other important ideas can be illustrated.

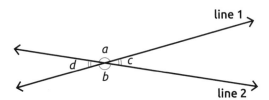

Angles *a* and *c* are **adjacent angles**—angles that share a common side and a common vertex. What other pairs of adjacent angles are in this diagram?

Angles *a* and *c* are also **supplementary angles**—two angles whose measures add up to 180°, so m∠*a* + m∠*c* = 180°. Notice that these angles together form a straight angle. Which other pairs of angles in the diagram are supplementary angles?

A different diagram is needed to show the next concept.

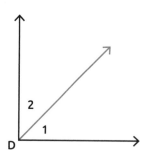

Let angle D be a right angle with a measure of 90°. Angle 1 and angle 2 compose, or make up, angle D. **Complementary angles** are two angles whose measures add up to 90°, so $m\angle 1 + m\angle 2 = 90°$.

Example 1

Solve this problem.

If $m\angle z = 51°$, find $m\angle w$.

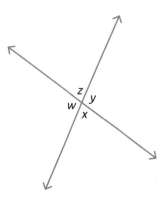

Angle z and angle w form a straight angle, so they are supplementary angles.

Using the definition of supplementary angles, $m\angle z + m\angle w = 180°$. Solve this algebraic equation:

$$m\angle z + m\angle w = 180°$$
$$51° + m\angle w = 180° \qquad \text{Substitute the known measure of angle } z.$$
$$51° - 51° + m\angle w = 180° - 51° \qquad \text{Subtract 51° from both sides.}$$
$$m\angle w = 129°$$

ANSWER: The measure of angle w is 129°.

Example 2

Using the diagram below, answer the question.

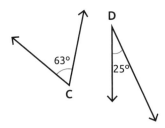

Are angles C and D supplementary, complementary, or neither?

Add the measures of the angles to check if the sum is 90°, 180°, or neither.

$63° + 25° = 88°$

ANSWER: The sum is less than 90°, so angles C and D are neither complementary nor supplementary.

Math Fact

The three angle measures in any triangle always add up to 180°.

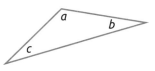

$m\angle a + m\angle b + m\angle c = 180°$

Complete the activities below to check your understanding of the lesson content. The Unit 5 Answer Key is on page 156.

Skills Practice

Fill in the blanks using the diagram below.

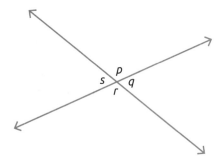

1. Angle *p* and angle _____ are vertical angles.

2. Angles *s* and _____ are supplementary angles.

3. Angle *p* forms a straight angle with angle _____.

4. Angle *s* has the same measure as angle _____.

5. If angle *r* measures 118°, angle *q* measures _____.

Use the diagram below to answer the questions.

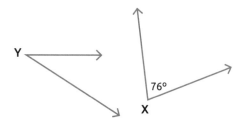

6. If angles X and Y are complementary angles, what is m∠Y? _____

7. Angle X and angle Y are both which type of angle?

    A. obtuse

    B. right

    C. acute

    D. straight

Perimeter is the total distance around a straight-sided figure. Your knowledge about squares, rectangles, triangles, and other shapes will be helpful to you in finding perimeter.

Perimeter problems are often found in real-world situations.

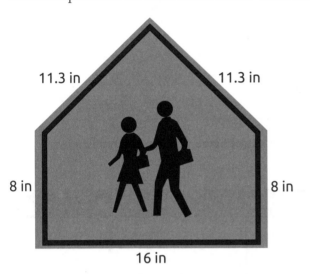

11.3 in   11.3 in

8 in   8 in

16 in

Math Fact

The number π, pronounced "pie," is a number whose decimal places never end or repeat. In calculations, 3.14 is often used as an approximation for π.

Example 1

A school crossing sign is in the shape of a trapezoid as shown. What is the perimeter of the sign?

To calculate the perimeter, find the distance around the figure. The lengths of all of the sides are given, so add them together: $11.3 + 11.3 + 8 + 16 + 8 = 56.6$.

ANSWER: The perimeter is 56.6 inches.

Circumference is the distance around a circle. To find circumference, you will use a special formula, or equation: $C = d\pi$, where d is the diameter of the circle. Or, since the diameter is two times the length of the radius, you could also use $C = 2r\pi$.

Example 2

Sometimes you are given the distance around a figure and asked to find another piece of information.

A car's tire travels 20,724 centimeters in 100 complete revolutions. How many centimeters is the tire's radius?

Since you need to find the radius, start with the formula $C = 2r\pi$ and solve for r. Use 3.14 to approximate π.

One complete revolution is the circumference of the tire, so 20,724 is 100 times the circumference of the tire. Translate from words to a math equation.

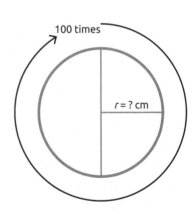

100 times

$r = ?$ cm

$20{,}724 = 100C$

$207.24 = C$ Use the value for C in the circumference formula and solve for r.

SOLVING PERIMETER AND AREA PROBLEMS

$$C = 2r\pi$$
$$207.24 = 2r\pi$$
$207.24 \approx 2 \times r \times 3.14$ Use ≈ instead of = because 3.14 is an approximation.
$207.24 \approx 6.28r$ Simplify
$\frac{207.24}{6.28} \approx \frac{6.28r}{6.28}$ Divide both sides by 6.28.
$33 \approx r$

ANSWER: The radius is about 33 centimeters.

Area is the amount of space inside a 2-dimensional figure and is measured in **square units**.

Shape	Formula for Area
Triangle	$A = \frac{1}{2}bh$ $= \frac{1}{2} \times$ base $\times$ height
Rectangle	$A = lw$ $=$ length $\times$ width
Trapezoid	$A = \frac{1}{2}(b_1 + b_2) \times h$ $= \frac{1}{2} \times$ sum of bases $\times$ height
Parallelogram	$A = bh$ $=$ base $\times$ height
Square	$A = s \times s$ $= s^2$ $=$ side squared
Circle	$A = \pi \times r \times r$ $= \pi r^2$ $= \pi \times$ radius squared

Math Fact

Notice that in the diagrams for area, a right angle mark is used to show that the height of a figure is measured using a line that is perpendicular to the base in a figure.

Example 3

A blacktop company needs to resurface a game court that is shaped like a semicircle, or half circle, on one end and a square on the other. If one side of the square part measures 22 feet, what is the approximate area of the game court?

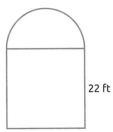

22 ft

You need to find the area of the square plus the area of the semicircle.

Notice that if you know one side of the square, you also know the length of the diameter of the circle because all sides of a square are equal. Remember that the diameter is twice as long as the radius. Use 3.14 for π.

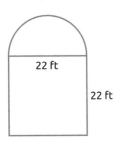

22 ft

22 ft

Skills Tip

Area is measured in square units. Finding area requires multiplication.

1 unit
1 unit ☐ Area = 1 square unit

2 units
3 units ⊞ Area = 6 square units

You will use two formulas: $A = s^2$ and $A = \pi r^2$.

Total Area = Area of square plus area of semicircle

Total Area = s^2 + $\frac{1}{2}(\pi r^2)$ Area of a semicircle is half the area of a circle.

Total Area = $22^2 + \frac{1}{2}(\pi \cdot 11^2)$ Substitute known values. If $d = 22$, $r = 11$.

Total Area ≈ $484 + \frac{1}{2}(379.94)$ Simplify.

Total Area ≈ 673.97

ANSWER: The area of the game court is about 673.97 square feet or 674 ft².

Complete the activities below to check your understanding of the lesson content. The Unit 5 Answer Key is on page 156.

Skills Practice

Solve the problems.

1. Connor is painting wooden boards cut into the shape of pine trees for a play. The top of each tree is in the shape of the triangle shown, where the sides marked *x* are both the same length and the third side measures 30.5 inches. If the perimeter of the triangular treetop is 115 inches, how long is each side marked *x*?

30.5 in

_____ inches

2. The track at Golden High School is made up of a rectangle with a semicircle on either end. Which of the following expressions could be used to find the distance that a runner would travel if she ran the total length of the outer perimeter of the track? Use 3.14 for π.

Hint: 2 semicircles equal 1 whole circle.

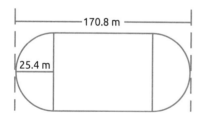

170.8 m

25.4 m

A. $170.8 \times 2 + 2 \times 25.4 \times 3.14$

B. $2(170.8 - 2 \times 25.4) + 4 \times 25.4 \times 3.14$

C. $170.8 - 2 \times 25.4 + 25.4^2 \times 3.14$

D. $2(170.8 - 2 \times 25.4) + 2 \times 25.4 \times 3.14$

3. A construction company is making a concrete patio shaped like a trapezoid. How many square feet is the area of the patio?

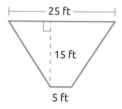

25 ft

15 ft

5 ft

_____ square feet

4. A square with side *s* is drawn inside a circle with radius *r*. If *s* = 1.7 millimeters and *r* = 1.2 millimeters, how many square millimeters is the area of the shaded region? Use 3.14 for π and round your answer to the nearest hundredth.

Hint: The shaded area is the difference between the areas of the two shapes.

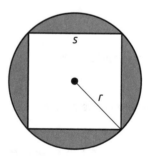

s

r

A. 0.88

B. 1.63

C. 4.65

D. 7.41

102

SOLVING SURFACE AREA AND VOLUME PROBLEMS

In this lesson, you will work with 3-dimensional (3-D) objects having length, width, and height.

The **surface area** of a 3-D object is the total space needed to cover the object. For example, when you wrap the box containing a birthday present in wrapping paper, the amount of paper needed to cover the box is the same as the surface area of the box. Surface area is measured in square units.

The **volume** of a 3-D object is the amount of space inside the object. For example, if you fill a bottle with water, the amount of water inside the bottle is the same as the volume of the bottle. Volume is measured in **cubic units**.

In this lesson, you will learn how to find the surface area and volume of the objects shown in the following table.

Skills Tip

Surface area is measured in square units because you are adding the 2-D areas of the flat or curved parts that make up the 3-D object.

Object	Surface Area	Volume
Cube	Since there are 6 equal sides to a cube, the surface area equals 6 times the area of 1 side, or face, of the square. All 6 sides have the same area. $SA = 6(s \times s)$ $SA = 6s^2$	Volume equals length times width times height (each measuring s). $V = s \times s \times s$ $V = s^3$
Rectangular solid	Surface area equals the sum of the areas of the top, bottom, front, back, and two sides of the rectangular solid. $SA = 2lw + 2wh + 2lh$	Volume equals length times width times height. $V = l \times w \times h$ $V = lwh$
Sphere	Use this surface area formula, where r is the radius of the sphere, or distance from the center of the sphere to a point on its surface. $SA = 4\pi r^2$	Use this volume formula. $V = \frac{4}{3}\pi r^3$
Cylinder	The surface area is the sum of the areas of the two circular ends and the rectangle that wraps around between them. The rectangle's length is equal to the circumference of one of the circles. The rectangle's width is equal to the height of the cylinder. $SA = 2\pi r^2 + 2\pi rh$	Volume is the area of the circular base times the height. $V = \pi r^2 h$

SOLVING SURFACE AREA AND VOLUME PROBLEMS

Example 1

On July 14, 2015, a spacecraft flew close to Pluto, sending back incredible photos. If Pluto's diameter is 2,274 kilometers, what is the surface area of Pluto? Use the surface area formula $A = 4\pi r^2$.

Pluto is approximately the shape of a sphere. You are given the diameter, but to use the formula, you need the radius. Remember that the radius is half the diameter, so the radius of Pluto measures $\frac{1}{2}(2,274)$, or 1,137 kilometers.

Use 3.14 to approximate the value of π.

$A = 4\pi r^2$

$A \approx 4(3.14)(1,137^2)$ Substitute known values.

$A \approx 16,237,178.64$

ANSWER: The surface area of Pluto is about 16,237,178.64 square kilometers.

Skills Tip

Volume is measured in cubic units. Finding volume requires multiplication.

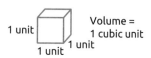

1 unit   Volume = 1 cubic unit

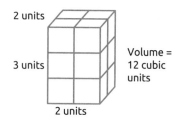

Volume = 12 cubic units

Example 2

Now, calculate the volume of Pluto. The formula for the volume of a sphere is $V = \frac{4}{3}\pi r^3$. Again, use 3.14 for π.

$V = \frac{4}{3}\pi r^3$

$V = \frac{4}{3}(3.14)(1,137)^3$ Substitute known values.

$V \approx 6,153,890,705$

ANSWER: The volume of Pluto is about 6,153,890,705 cubic kilometers.

Note: You would read this number as "six billion, one hundred fifty-three million, eight hundred ninety thousand, seven hundred five."

Example 3

Sharon owns the barrel shown in the figure on the right. The barrel has an opening at the top so that rainwater from a gutter can be collected. If the radius r of the circular base is 2.5 feet and the barrel can hold approximately 75 cubic feet of rainwater, about how tall is the barrel? Use 3.14 for π and round to the nearest whole number.

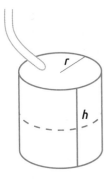

Use the formula for the volume of a cylinder: $V = \pi r^2 h$.

$V = \pi r^2 h$

$75 \approx 3.14(2.5^2)h$ Substitute known values.

$\dfrac{75}{3.14(2.5^2)} \approx h$ Solve for h.

$3.8216 \approx h$

$4 \approx h$ Round to the nearest whole number.

ANSWER: The rain barrel is about 4 feet tall.

Complete the activities below to check your understanding of the lesson content. The Unit 5 Answer Key is on page 156.

Skills Practice

Solve the problems.

1. A cardboard paper towel tube is 11 inches long. The diameter of the open ends of the tube measures 1.5 inches across. How many square inches of cardboard is needed to make the paper towel tube? Use 3.14 for π and round your answer to the nearest hundredth, if necessary.

    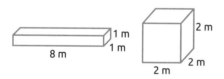

    A. 19.43

    B. 38.86

    C. 51.81

    D. 55.34

2. Which of the following statements about the relationship between the rectangular solid and cube shown is correct?

    A. The surface area of the rectangular solid is less than the surface area of the cube, but their volumes are the same.

    B. The surface area of the rectangular solid is greater than the surface area of the cube, but their volumes are the same.

    C. The volume of the box is greater than the volume of the cube, but their surface areas are the same.

    D. The volume of the box is less than the volume of the cube, but their surface areas are the same.

3. Tammy wants to stain a wooden jewelry box that is 10 inches long by 10 inches wide by 3 inches high. She needs to stain all of the sides and the top, but not the bottom of the box because it won't been seen. How many square inches is the surface area of the wood that needs to be stained?

    _____ square inches

4. A grain silo is in the shape of a cylinder domed with half of a sphere as shown. Choose the correct expression for finding the exact volume of grain that the silo holds if it can be filled to the very top of the half-sphere dome.

    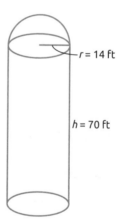

    A. $\pi(14^2)(70) + \frac{1}{2} \times \frac{4}{3}\pi(14^3)$

    B. $\pi(28^2)(70) + \frac{1}{2} \times \frac{4}{3}\pi(28^3)$

    C. $\pi(14^2)(70) + \frac{4}{3}\pi(14^3)$

    D. $\pi(28^2)(70) + \frac{4}{3}\pi(28^3)$

Answer the questions based on the content covered in this unit. The Unit 5 Answer Key is on page 156.

Solve the problems.

1. Which of the following statements is true?

   A. A parallelogram is also a rectangle.

   B. A rectangle is also a parallelogram.

   C. All quadrilaterals are also squares.

   D. All parallelograms are also rhombuses.

2. Which geometric shapes have sides that are always equal in length?

   A. triangles and pentagons

   B. rhombuses and squares

   C. rectangles and parallelograms

   D. rectangles and trapezoids

Give the coordinates of the indicated points.

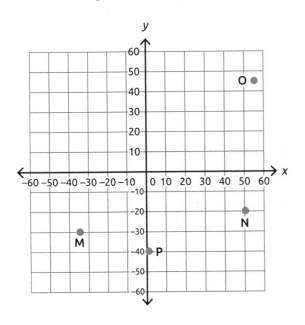

3. Point M (_____, _____)

4. Point N (_____, _____)

5. Point O (_____, _____)

6. Point P (_____, _____)

Plot the points on the coordinate plane.

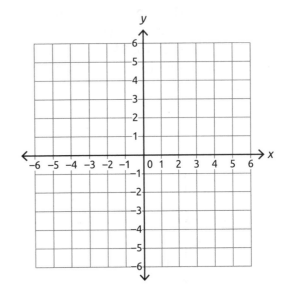

7. Point R at (4, −6)

8. Point S at (1.5, 3.5)

9. Point T at (−5, −1)

10. Point U at (0, −2.5)

Solve the problem.

11. The coordinate grid below is used to map out a recreational lake that has an island in the middle. On this map, 1 unit equals 1 kilometer. A seafood restaurant is located at point R. If a boater sails from the restaurant due east for 5 kilometers, then due south for 7.5 kilometers to his house, what are the coordinates of his house?

_____

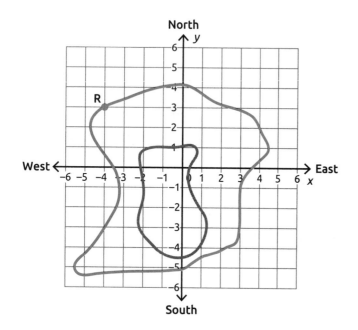

Determine whether the following statements are true or false using the diagram below. None of the angles measure 90°.

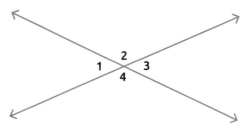

12. m∠1 + m∠3 = 180° _____

13. m∠2 = m∠4 _____

14. If m∠3 = 48°, then m∠2 = 132°. _____

15. ∠4 is adjacent to ∠2. _____

Solve the problems.

16. In rectangle MNOP, angles P and M are _____.

A. supplementary angles

B. acute angles

C. complementary angles

D. obtuse angles

17. A small oil painting is shown in a frame. The length of the painting is 8 inches longer than its width. If the perimeter of the painting is 24 inches, how many inches is the length of the painting?

A. 2

B. 4

C. 8

D. 10

18. Which of the following statements is correct?

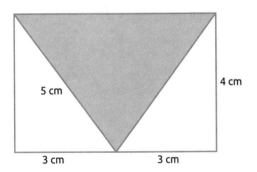

5 cm

4 cm

3 cm 3 cm

A. The area of the shaded triangle is half as large as the area of the entire rectangle.

B. The area of the shaded triangle is one-third as large as the area of the entire rectangle.

C. The area of the shaded triangle is two-thirds as large as the area of the entire rectangle.

D. There is not enough information to compare the areas.

19. A sketch of a gold hoop earring is drawn using two circles. The diameter of the larger circle is 2.2 inches, and the diameter of the smaller circle is 1.6 inches. How many square inches is the gold area that makes up the hoop? Use 3.14 for π and round your answer to the nearest hundredth.

A. 5.81 in²

B. 2.36 in²

C. 1.79 in²

D. 0.57 in²

20. A puzzle cube measures 5.7 centimeters on each side. If the colored stickers cover 96% of the surface, how many square centimeters of the cube is covered by stickers? Round your answer to the nearest tenth.

A. 179.2 cm²

B. 187.1 cm²

C. 192.8 cm²

D. 194.9 cm²

21. A 3-D printer has created a plastic object that is a cube with a hollowed-out spherical core. The diameter of the sphere is 8 millimeters less than the length of a side of the cube. How many cubic millimeters of plastic are used to create the object? Use 3.14 for π and round to the nearest hundredth, if necessary.

Hint: The formula for the volume of a sphere is $V = \frac{4}{3}\pi r^3$.

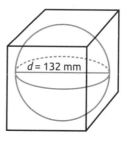

d = 132 mm

A. 1,307,973.33 mm³

B. 1,540,350.08 mm³

C. 2,744,000.00 mm³

D. 3,947,649.92 mm³

Measurement and Data

The state of California experienced an extreme drought. Almond growers in particular got complaints that they were using too much water.

A bar graph is a way to display facts and other information. The bar graph below compares the water use among various crops grown in California.

California's Annual Agricultural Water Use (in Million Acre-Feet)

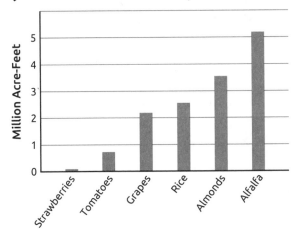

In this case, the data, or pieces of information, that were measured and displayed were the numbers of acre-feet of water used.

In this unit, you will learn how to take measurements in several measurement units as well as how to represent data visually using graphs such as this one.

Unit 6 Lesson 1

MEASURING AND ESTIMATING LENGTHS IN STANDARD UNITS

A **ruler** is a straight object that usually is marked at regular intervals and is used to measure the length of objects or distances.

In the **U.S. customary system**, length or distance is measured (from shortest unit to longest) in inches (in), feet (ft), yards (yd), and miles (mi). A standard ruler is 12 inches and is marked with inches and fractions of an inch. Longer rulers and measuring tapes are also available. A yardstick is 1 yard, or 36 inches, and is marked with feet and inches. A measuring tape is typically 16 feet, or 192 inches. If a measuring tape were available to measure a mile, it would have to have 63,360 inches marked on it!

Sometimes length or distance is measured using the **metric system** of millimeters (mm), centimeters (cm), meters (m), and kilometers (km). The metric system is a convenient base ten system. Each unit is 10 times smaller than the next larger unit. A standard metric ruler is 30 centimeters and is marked with centimeters and millimeters.

MEASURING AND ESTIMATING LENGTHS IN STANDARD UNITS

Real-World Connection

A measuring wheel can be used to measure long distances in various units.

The ruler pictured here shows centimeters on the top and inches on the bottom.

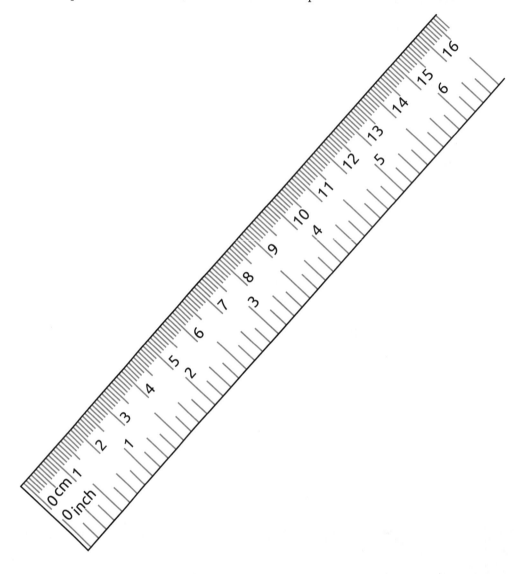

Example 1

What is the length of the pen shown here?

To use a ruler to measure the pen in inches, line up the end of the pen at 0 inches as shown.

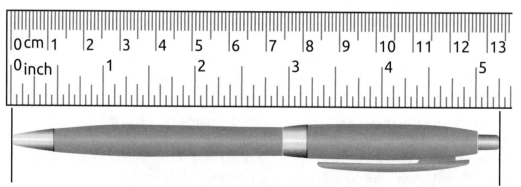

The ruler has 16 tick marks from one inch to the next. Notice that the longest marks on the ruler are the whole inches and the $\frac{1}{2}$ inches. The next longest marks are the $\frac{1}{4}$ inches. The ruler also has smaller marks for $\frac{1}{8}$ and $\frac{1}{16}$ inches.

ANSWER: The pen extends past the 5-inch mark. Each of the smallest tick marks is $\frac{1}{16}$ inch, so 4 of them would be $\frac{4}{16}$. The pen measures $5\frac{4}{16}$, or $5\frac{1}{4}$ inches long. Note that you could have used the $\frac{1}{4}$-inch, $\frac{1}{8}$-inch, or $\frac{1}{16}$-inch marks after 5 to determine that the pen is $5\frac{1}{4}$ inches long.

Example 2

What is the length of the pen in centimeters?

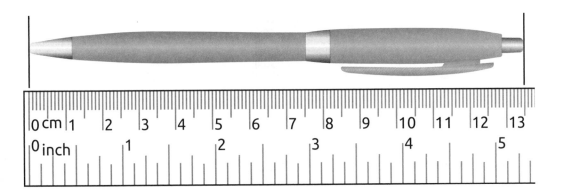

The ruler has exactly 10 tick marks from one centimeter to the next. Each of these tick marks is 1 millimeter.

Notice that the longest marks on the ruler are the whole centimeters (cm), the next longest marks are the 0.5 centimeters, and the smallest marks are the millimeters (mm).

ANSWER: The pen extends past the 13-centimeter mark. Each of the smallest tick marks is 1 mm, so 5 of them is 5 mm. The pen is 13 cm plus 5 mm long. However, you need to give your answer in centimeters.

1 millimeter is 0.1 centimeter, so 13 cm + 5 mm = 13 cm + 0.5 cm = 13.5 cm.

You could also look at it this way—in centimeters, the pen extends halfway past the 13-cm mark. So, the pen is 13.5 centimeters long.

Complete the activities below to check your understanding of the lesson content. The Unit 6 Answer Key is on page 157.

Skills Practice

Answer the questions.

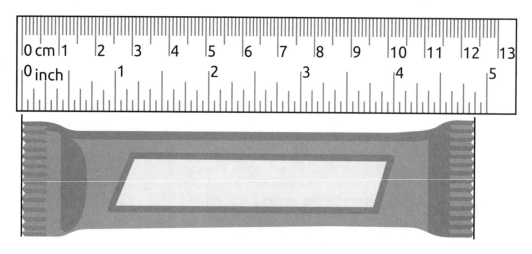

1. What is the length of the candy bar in inches? _____

2. What is the length of the candy bar in centimeters? _____

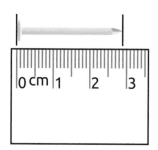

3. What is the length of the nail in centimeters? _____

4. What is the length of the nail in millimeters? _____

SOLVING PROBLEMS INVOLVING MEASUREMENT

In this lesson, you will solve problems that involve units of time, liquid volume, and mass.

If someone asks, "What time is it?" you probably check your cell phone's digital display. For instance, it might read, "12:45" where 12 is the hour and 45 is the number of minutes past the hour. If it is after noon, you would say it is 12:45 PM. If it is after midnight, it is 12:45 AM.

Example 1

What is the time shown on this clock?

This system for marking time uses a **12-hour clock**. First, look at the short **hour** hand. It points between 7 and 8, so the hour part of the time is 7. From one number to the next are five tick marks, or five minutes. You can see a total of 60 tick marks (12 × 5) around the clock, marking 60 minutes in each hour.

Since the longer **minute** hand points to the 3, then 15 minutes (3 × 5) have passed since 7 o'clock.

ANSWER: This clock reads 7:15. It is unclear whether it is AM or PM.

Notice that the minute hand has swept $\frac{1}{4}$ of the way around the clock. This means you could also read the time as "a quarter after 7."

Vocabulary Tip

The abbreviation *AM* comes from the Latin phrase "ante meridiem," before midday, and *PM* comes from "post meridiem," after midday.

Example 2

You may already be familiar with the liquid measurement of liter if you have poured soda from a two-liter bottle.

A **liter** (l) is a metric unit of capacity of a gas or a liquid. A **milliliter** (ml) is 0.001 of a liter, so 1000 milliliters = 1 liter.

A chemist has a glass beaker that measures liquids. The number 1000 at the top right is the 1000 milliliter mark. The liquid in the beaker is up to the mark, so the beaker contains 1000 milliliters, or 1 liter, of liquid.

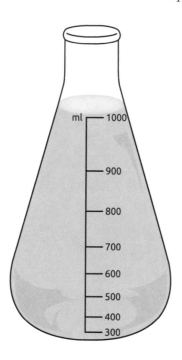

Math Fact

One liter has the same capacity as a cube with sides measuring 10 centimeters.

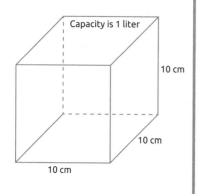

If the chemist pours out one-fourth of the liquid, how many milliliters remain? How many liters remain?

ANSWER: The beaker contains 1000 ml of liquid.

one-fourth of 1000 ml = $\frac{1}{4} \times 1000$ ml = 250 ml

1000 ml − 250 ml = 750 ml   Subtract the liquid that is poured out.

So, 750 ml remain in the beaker.

Since 1 l = 1000 ml, divide 750 by 1000 to find out how many liters remain.

To divide by 1000, move the decimal point in a number three places to the left. 750 has an implied decimal to the right of the ones place.

750. ÷ 1000 = .750

It makes sense that less than 1 liter would be left in the beaker because the chemist started with 1 liter of liquid and then poured some out.

Example 3

The **mass** of an object is the amount of matter, or "stuff," that makes up the object. While we use milliliters and liters to measure the volume or capacity of a gas or a liquid, we use grams and kilograms to measure the mass of a solid.

A **gram** (g) is a metric unit of capacity of the mass of a solid. The mass of one paper clip is about 1 gram. A **kilogram** (kg) is equal to 1000 grams. The mass of a textbook is about 1 kilogram.

> **BE CAREFUL!**
>
> Mass and weight are often confused. The weight of something depends on the pull of gravity, but the mass of something doesn't. For example, the average adult weighs 70 kg on Earth but would only weigh about 11.7 kg ($\frac{1}{6}$ of 70 kg) on the moon because the moon's force of gravity is $\frac{5}{6}$ less. However, the mass (also 70 kg) would remain the same regardless of the location because the amount of matter stays the same.

Describe how to balance the scale shown here.

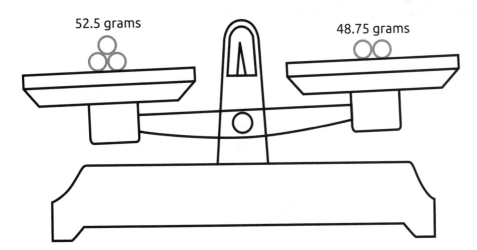

52.5 grams   48.75 grams

ANSWER: Add mass to the right side to balance the scale. Subtract to find how many grams to add to the right side: 52.5g − 48.75 g = 3.75 g.

> **Vocabulary Tip**
>
> A **milligram** (mg) equals 0.001 of a gram. Milligrams are used to measure very small amounts of mass, like that of a grain of salt or a tiny insect.

Complete the activities below to check your understanding of the lesson content. The Unit 6 Answer Key is on page 157.

Skills Practice

Answer the questions.

1. If all of the beakers shown here were filled to their top mark and then combined in a single empty container, how many milliliters of liquid would be in the container? How many liters of liquid would be in the container?

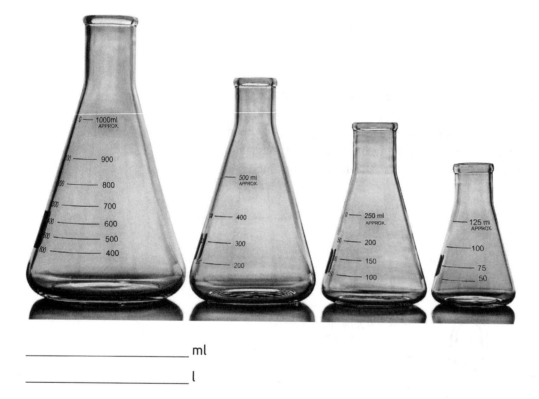

_____ ml

_____ l

2. Look at the time on the clock shown here. What would be the time 42 minutes later than the time shown?

   A. 1:42
   B. 1:50
   C. 2:00
   D. 2:10

3. The mass of Object Y is 10 more than 4 times the mass of Object X. Object Y has a mass of 330 kg. Find the mass of Object X in kilograms and in grams.

   A. 80 kg; 80,000 g
   B. 80 kg; 8000 g
   C. 85 kg; 85,000 g
   D. 85 kg; 8500 g

CONVERTING MEASUREMENT UNITS

In this lesson, you will continue to work with measurement units and learn more about changing from one unit to another.

Because it uses a base ten system, the metric system allows you to convert between units simply by moving the decimal point.

Study the following table.

kilo	hecto	deca	meter (m) liter (l) gram (g)	deci	centi	milli
			1.0			
Divide by 1,000	Divide by 100	Divide by 10		Multiply by 10	Multiply by 100	Multiply by 1,000
Move left 3	Move left 2	Move left 1		Move right 1	Move right 2	Move right 3
0.001	0.01	0.1		10	100	1,000

At the center of the table, notice that meter, liter, and gram are the basic metric units introduced in previous lessons.

If you are given a measurement in one of these units and need to convert to a different unit, you would move the decimal point to the left or to the right the number of times listed in the table.

Example 1

Convert 0.82 liter to centiliters.

ANSWER: 0.82 liter equals 82 centiliters because you would move the decimal point 2 places to the right.

Example 2

Convert 22,875 grams to hectograms.

ANSWER: 22,875 grams equals 228.75 hectograms because you would move the decimal point 2 places to the left.

Skills Tip

Remember that every whole number has an implied decimal point to the right of the ones place.

You are not, however, restricted to starting with the basic units.

- To convert any metric unit to a *smaller* unit, move the decimal to the *right* the appropriate number of times, using the chart to guide you.

- To convert any metric unit to a *larger* unit, move the decimal to the *left* the appropriate number of times, using the chart to guide you.

Example 3

Convert 132 milligrams to kilograms.

Write 132.0 mg in the milli column. The kilo column is 6 places to the left of the milli column, so move the decimal point 6 places to the left.

kilo	hecto	deca	meter (m) liter (l) gram (g)	deci	centi	milli
0.000132 kg						132.0 mg

ANSWER: 132 mg = 0.000132 kg

Example 4

Convert 0.418 kilometers to meters.

Write 0.418 km in the kilo column. The meter column is 3 places to the right of the kilo column, so move the decimal point 3 places to the right.

kilo	hecto	deca	meter (m) liter (l) gram (g)	deci	centi	milli
0.418 km			418 m			

ANSWER: 0.418 km = 418 m

Changing from one unit to another in the U.S. customary system of measurement is not as convenient. Take a look at the equivalencies among customary length units.

- To convert from a larger unit to a smaller unit, *multiply*.
- To convert from a smaller unit to a larger unit, *divide*.

Example 5

Convert 48 inches to feet.

Use the conversion equation 12 inches = 1 foot. Since an inch is smaller than a foot, divide.

$48 \div 12 = 4$

ANSWER: 48 in = 4 ft

Example 6

Convert 17 yards to inches.

Use the conversion equation 36 inches = 1 yard. Since a yard is larger than an inch, multiply.

$17 \times 36 = 612$

ANSWER: 17 yd = 612 in

Math Fact

It is also possible to convert between U.S. customary and metric units.

For example, 1 in = 2.54 cm.

Complete the activities below to check your understanding of the lesson content. The Unit 6 Answer Key is on page 157.

Skills Practice

Convert the following measurements.

1. Convert 0.943 meter to decimeters.

   _____ decimeters

2. Convert 175 kilometers to millimeters.

   A. 0.000175 mm

   B. 0.175 mm

   C. 175,000 mm

   D. 175,000,000 mm

3. Convert 0.8 miles to feet. Use 1 mi = 5,280 ft.

   A. 28.8 ft

   B. 4,224 ft

   C. 5,280 ft

   D. 6,600 ft

4. Convert 7,236 inches to yards. Use 1 yd = 36 in.

   A. 201 yd

   B. 603 yd

   C. 86,832 yd

   D. 260,496 yd

In this lesson, you will learn how to use graphs as visual aids for making sense of data. **Data** are measurements, facts, or other pieces of information that have been collected in some way.

A **bar graph** is a graph that makes it easier to compare data by representing values using bars of various lengths on the same scales. Bar graphs are used to show data that have been put into categories.

A **line graph** is a graph that shows data points connected by line segments. Line graphs make it easier to see trends over time.

Vocabulary Tip

A **histogram** is a type of bar graph that often uses intervals of values on the horizontal axis. For example, 0–9 and 10–19 are intervals. The bars are usually drawn without gaps between them.

Example 1

Theo collected tadpoles from several ponds and measured each one's length to the nearest quarter of an inch. He created this bar graph of the results.

Tadpole Survey Results

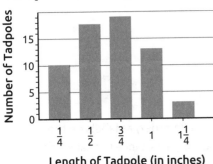

What was the most common length of the tadpoles Theo measured?

ANSWER: The tallest bar shows that 19 tadpoles measured $\frac{3}{4}$ inch, so the most common length was $\frac{3}{4}$ inch.

How many tadpoles did Theo measure?

ANSWER: Add the number of tadpoles for each length. Find the height of each bar, then add the values: $10 + 18 + 19 + 13 + 3 = 63$. Theo measured 63 tadpoles in all.

What percent of the tadpoles measured $1\frac{1}{4}$ inches long? Round to the nearest percent.

ANSWER: Use the percent equation: percent = part ÷ whole. Three of the 63 tadpoles Theo measured were $1\frac{1}{4}$ inches long, so part = 3 and whole = 63.

percent = part ÷ whole

percent = 3 ÷ 63

percent ≈ 0.047619 ≈ 5%

About 5% of the tadpoles were $1\frac{1}{4}$ inches long.

Skills Tip

Bar graphs and line graphs should always have labels and scales on both axes and a title at the top that describes what is being represented by the graph.

REPRESENTING AND INTERPRETING DATA

Example 2

Xing-Hai is a quality-assurance inspector of a type of car part being made at a factory. The car part must weigh either 0.32 or 0.33 kg in order to pass inspection.

Eighteen parts are randomly weighed to the nearest hundredth of a kilogram and recorded as shown.

0.32   0.33   0.31   0.31   0.31   0.31

0.34   0.32   0.34   0.33   0.33   0.32

0.32   0.33   0.33   0.33   0.32   0.33

Create a bar graph that displays the number of parts for each weight listed in the sample of 18 parts.

ANSWER: First, take a tally of each weight in the list.

Weight in kg	Tally
0.31	IIII
0.32	IIII
0.33	IIII II
0.34	II

Then, create the bar graph. Use the four different weights (in order) as labels on the horizontal axis. The vertical axis should show the number of parts. Since the tallies are between 2 (least) and 7 (most), a scale from 0 to 8 would work well for the vertical axis.

Number of Car Parts of Certain Weights in Sample

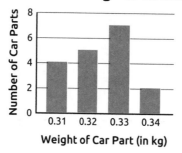

What is the ratio of parts that fail inspection to parts that pass?

ANSWER: Recall that a part must weigh either 0.32 kg or 0.33 kg in order to pass inspection.

Write a ratio (fraction) comparing the number of parts that do not weigh those amounts to the parts that do.

$$\frac{\text{parts that fail (0.31 kg or 0.34 kg)}}{\text{parts that pass (0.32 kg or 0.33 kg)}} = \frac{4+2}{5+7} = \frac{6}{12} = \frac{1}{2}$$

The ratio of parts that fail to parts that pass inspection is 1:2.

Example 3

A lab assistant reads the graph to determine how much of two different chemicals she needs to add to a solution each week for five weeks.

Amount of Chemicals A & B in Solution

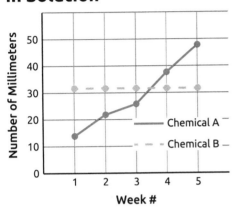

Which week calls for the lab assistant to put more of Chemical A than Chemical B into the solution for the first time?

ANSWER: The solid line represents the amount of Chemical A that gets added to the solution. Since the solid line crosses and is then higher than the dotted line at Week 4, the lab assistant will add more of Chemical A to the solution for the first time at Week 4.

Describe the trend in the amount of Chemical A that is being added over the five weeks.

ANSWER: The solid line representing Chemical A rises from left to right, so the trend is increasing amounts of Chemical A over the five weeks.

Describe the trend in the amount of Chemical B that is being added.

ANSWER: The dotted line representing Chemical B is flat, so the trend is for the amount of Chemical B being added to remain constant, or steady, at 32 ml over the five weeks.

Vocabulary Tip

A line that rises from left to right has **positive slope**, one that falls from left to right has **negative slope**, and one that is horizontal (flat) has **zero slope**.

Complete the activities below to check your understanding of the lesson content. The Unit 6 Answer Key is on page 157.

Skills Practice

Answer the following questions based on the bar graph.

A national park has four kinds of hooved animals that graze there. The graph shows the percent of each type of these animals.

Percent of Different Animals at Park

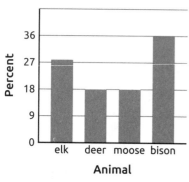

1. The percentage of elk at the park is 28%. What is the difference between the percentage of elk at the park and the percentage of moose?

   _____

2. If there are 400 hooved animals at the park in all, how many of them are bison?

   _____

Answer the following questions based on the line graph.

Counts are taken several times per day of the number of people attending Beach X.

People at Beach X

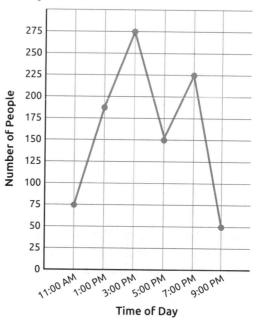

3. At which time of day are the most people at Beach X?

   _____

4. What is the trend in attendance between 11:00 AM and 3:00 PM?

   A. Attendance increases.

   B. Attendance decreases.

   C. Attendance stays the same.

   D. Attendance increases and then decreases.

Answer the questions based on the content covered in this unit. The Unit 6 Answer Key is on page 157.

Answer the questions.

1. What is the length of the leaf in inches? In centimeters?

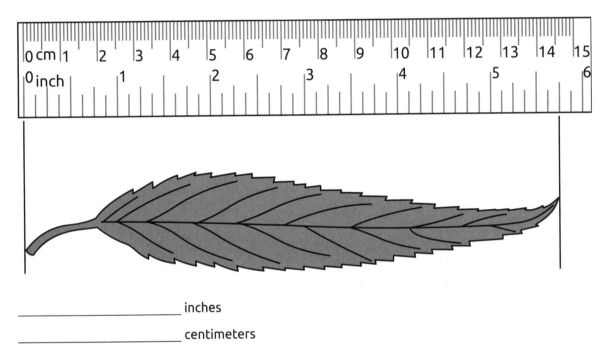

_____ inches

_____ centimeters

2. What time is shown on the clock?

_____

3. The glass container measures liquids in milliliters. If the container is filled with water to 250 ml each time, how many containers are needed to fill a 2-liter bottle?

A. 4

B. 6

C. 8

D. 10

4. Francine has a crate of 78 rpm records from the early 1950s in her basement. If there are 19 records in the crate and each one weighs 220 grams, what is the weight of the records in kilograms?

_____ kg

Solve the problems.

5. Convert 402,678 centimeters to kilometers.

_____ km

6. Convert 18 yards to feet.

   A. 1.5 ft

   B. 6 ft

   C. 54 ft

   D. 216 ft

Answer the questions based on the bar graph.

The number of hits in seven baseball games is shown on the graph.

Hits Made During Seven Games

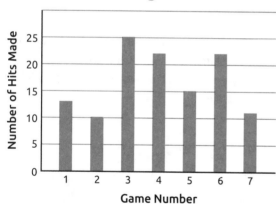

7. What was the fewest number of hits made in a game?

_____ hits

8. Which two games had the same number of hits?

_____

Answer the questions based on the line graph.

The height of a certain plant was recorded on the first of the month from April to October, and its growth was then graphed.

Recorded Height of Plant

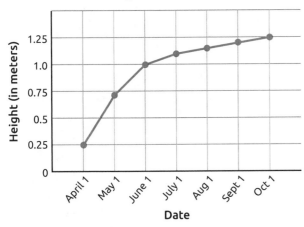

9. How much taller was the plant on October 1 than on April 1?

   _____

10. Describe the trend in the growth of the plant.

   _____

   _____

Visitors from around the world come to see Old Faithful erupt.

Geysertimes.org records the time between eruptions. On July 27, 2015, of the last 100 recorded time intervals, the shortest wait time between eruptions was 1 hour 1 minute, and the longest wait time was 9 hours 19 minutes.

The mean time between intervals was 2 hours 3 minutes, while the median time was 1 hour 33 minutes. Both *mean* and *median* are ways to measure the center value of a set of data.

Geysertimes.org asked a statistical question, "What is the wait time between eruptions?" In this unit, you will learn that the set of data collected to answer a statistical question has important features. These features help us see the "big picture."

Statistics is the science of collecting data in order to analyze, interpret, and display it in helpful ways. One important statistical feature for a set of data is its **measure of center**, or the value that represents the middle of a set of data. You have three ways to find the measure of center.

Mean	Divide the total (sum) of all of the values in a set by the number of values. The mean is most useful when there are few or no extreme values in the set, called **outliers**. Outliers will cause the mean to be distorted.
Median	List the values from least to greatest, then locate the value in the middle. If you have an even number of values, find the mean of the two values in the middle. The median is helpful when there are some outliers because these extreme values do not affect the median.
Mode	Count the number of times each value appears in a set. The value that occurs most often is the mode.

To determine how different data points are from one another, find the **range**—the difference between the least value and the greatest value.

Example 1

Find the mean, median, and mode of the following set of test scores: 78, 82, 92, 79, 89, 84.

ANSWER: To find the mean, add the scores and divide by the number of scores, 6.

$$\frac{78 + 82 + 92 + 79 + 89 + 84}{6} = 84 \qquad \text{The mean is 84.}$$

To find the median, put the scores in order, and then average the two middle scores since it is an even number of scores.

$$78 \quad 79 \quad 82 \quad \uparrow \quad 84 \quad 89 \quad 92$$

The mean of 82 and 84 is $\frac{82 + 84}{2} = 83$. The median is 83.

Notice that the mean and the median are very close. Therefore, either of them would be equally helpful to describe the center of the data.

Since no score occurs more than any other score, there is no mode.

UNDERSTANDING STATISTICAL VARIABILITY

Math Fact

A set of data may have one mode, more than one mode, or no mode.

Example 2

Recall that of the last 100 recorded time intervals between Old Faithful eruptions, the shortest wait time between eruptions was 1 hour 1 minute, and the longest was 9 hours 19 minutes. There is a big difference between these two values!

The range is the longest time interval minus the shortest: 9h 19m – 1h 1m = 8h 18m.

Take a look at the actual data. Notice that most of the time intervals are about 1h 30 m (1.5 hours) long. The highlighted values are outliers because they are much longer than most of the others.

1h 40m	3h 6m	1h 32m	1h 31m	1h 29m	1h 38m	4h 30m	1h 40m	1h 41m	1h 28m
1h 25m	1h 27m	1h 32m	1h 57m	1h 45m	1h 32m	1h 21m	3h 5m	1h 32m	1h 18m
1h 37m	1h 31m	1h 25m	1h 42m	1h 28m	1h 34m	1h 43m	1h 29m	1h 27m	1h 40m
1h 31m	1h 26m	1h 33m	1h 42m	1h 32m	1h 25m	1h 23m	1h 34m	1h 24m	1h 29m
3h 22m	1h 34m	1h 37m	8h 59m	1h 32m	1h 32m	1h 33m	1h 18m	4h 28m	1h 37m
1h 23m	1h 40m	1h 34m	1h 32m	1h 30m	1h 23m	4h 13m	1h 40m	1h 34m	1h 36m
1h 53m	2h 53m	1h 28m	1h 28m	1h 28m	1h 43m	4h 52m	1h 24m	1h 33m	1h 34m
1h 29m	1h 25m	1h 51m	1h 31m	1h 42m	1h 30m	1h 34m	1h 43m	1h 24m	1h 26m
1h 48m	1h 40m	1h 28m	2h 55m	9h 19m	6h 25m	1h 35m	1h 2m	1h 31m	7h 33m
1h 25m	8h 0m	1h 36m	1h 35m	1h 36m	1h 1m	1h 34m	1h 40m	1h 28m	1h 30m

Vocabulary Tip

Another word for mean is **average.**

The mean time between intervals was 2 hours 3 minutes, while the median time was 1 hour 33 minutes. The mean is higher than the median because of the extremely high outliers—the times when it took the geyser much longer between eruptions. The median is a better number to describe the center of this set.

Complete the activities below to check your understanding of the lesson content. The Unit 7 Answer Key is on page 157.

Skills Practice

Answer the questions based on the data in the table.

The outdoor temperature, in degrees Fahrenheit, is recorded at noon at a middle school for seven consecutive days.

Date	°F
July 8	83°
July 9	87°
July 10	92°
July 11	95°
July 12	82°
July 13	83°
July 14	81°

1. Find the mean, median, and mode of the data. Round your answers to the nearest tenth, if necessary.

   Mean: _____

   Median: _____

   Mode: _____

2. Which statement about the measures of center is true?

   A. The mode is the worst measure of center because most of the temperatures are higher than the mode.

   B. The median is the worst measure of center because outliers make the median too low.

   C. The mean is the worst measure of center because outliers make the mean too high.

   D. The range is the best measure of center because it uses the highest and lowest temperatures.

3. What is the range of the data?

   A. 95°F

   B. 81°F

   C. 14°F

   D. 11°F

DESCRIBING DISTRIBUTIONS

In the previous lesson, you learned how to calculate the range of a set of data values as well as the ways of finding the mean (average), median (middle), and mode (most frequent value).

In this lesson, you will learn about the ways that data are **distributed**, or spread out, when graphed as a histogram and how this relates to the measures of center.

Recall that a histogram is a type of bar graph that often uses intervals of values on the horizontal axis. The bars in a histogram are next to each other because the data are usually **continuous**, meaning a bar can have any value within an interval of values.

A histogram can reveal how data is distributed. Take a look at these four types of distribution.

Spread out more on the left

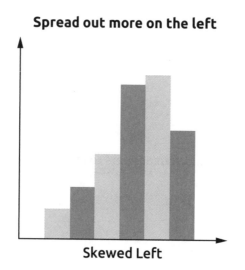

Skewed Left

Spread out more on the right

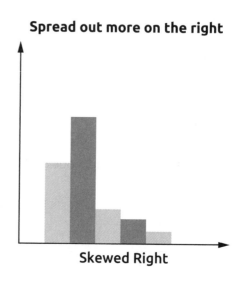

Skewed Right

No visible pattern

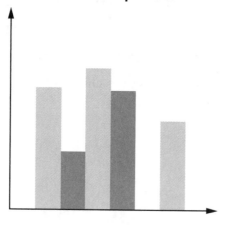

Symmetrical and balanced

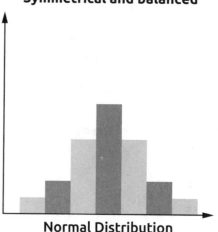

Normal Distribution

Vocabulary Tip

The normal distribution is sometimes called a **bell curve** because of its shape.

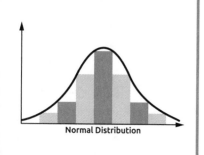

Normal Distribution

A *perfectly* **normal distribution** is:

- where the mean, median, and mode are all equal and are located in the center of the distribution.

- symmetrical—half of the values are less than the mean, and half are greater than the mean.

- often found for data sets such as a person's height, intelligence, and blood pressure.

Example 1

In an aerobics class, the teacher has students find their resting heart rates before they begin. The number of beats per minute for each of the 36 students is shown.

66	71	65	69	77	73	81	92	72	69	79	88
76	71	81	72	59	75	78	110	84	70	68	81
70	80	78	76	71	75	80	68	64	71	90	75

Draw a histogram of the data using the intervals 50–59, 60–69, 70–79, 80–89, 90–99, 100–109 and 110–119.

ANSWER: First, take a tally of each resting heart rate by interval.

Interval	Tally
50–59	l
60–69	llll ll
70–79	llll llll llll lll
80–89	llll ll
90–99	ll
100–109	
110–119	l

Then, create the histogram where the seven intervals are on the horizontal axis and the number of data points (tallies) is shown on the vertical axis, labeled "Frequency."

Resting Heart Rates in Aerobics Class

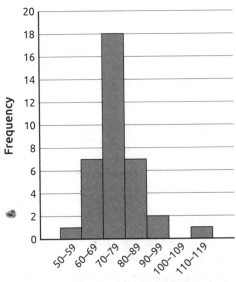

Notice that the intervals are all the same size (10 beats per minute) and that they cover all of the data points so that none are left out of an interval.

DESCRIBING DISTRIBUTIONS

Since the number of tallies is between 1 (least) and 18 (most), a scale from 0 to 20 would work well for the vertical axis.

What is the shape of the distribution?

ANSWER: At first, you might think that the histogram appears to be fairly normally distributed, but notice the outlier in the 110–119 interval. The outlier in the data set is 110 bpm.

Since the data is drawn out to the right by this outlier, the distribution is skewed right.

In which interval would you expect to find the mean? The median? The mode?

ANSWER: Because of the outlier at 110 bpm, you would expect the mean to be higher than the median or the mode, but because most of the data seems to be in the 70–79 bpm interval, you might guess that all three measures are within that interval.

CHECK: Find the mean, median, and mode of the data.

To find the median, arrange the values from least to greatest, and then average the middle two values, 75 bpm and 75 bpm. The median is 75 bpm.

To find the mean, add all of the values (the sum is 2,725), and then divide by 36. Rounded to the nearest whole, the mean is 76 bpm, slightly right of center of the 70–79 interval. The mean is higher than the median.

The mode, or the data value that occurs most often, is 71.

In this case, all three measures of center do fall within the expected interval of 70–79.

Example 2

> ### Vocabulary Tip
>
> The number of times a data value occurs in a data set is called its frequency.

Recall from a previous lesson that GeyserTimes.org found that the mean of 100 time intervals between Old Faithful eruptions was 2 hours 3 minutes, while the median was 1 hour 33 minutes.

What can this information tell you about the distribution of the data?

Since the mean is larger than the median, you can assert that the distribution is not perfectly normal.

You might guess that the distribution is skewed right, but you would have to graph the data to make sure that is correct.

> ### BE CAREFUL!
>
> When reading a histogram, don't assume that the mode is the value that is in the middle of the tallest bar. Because the data has been graphed using intervals of values, you must go back to tally the data points themselves to determine the mode.

LESSON REVIEW

Complete the activities below to check your understanding of the lesson content. The Unit 7 Answer Key is on page 157.

Skills Practice

Answer the questions based on the histogram.

A physics teacher created this histogram of the scores her students earned on the latest quiz.

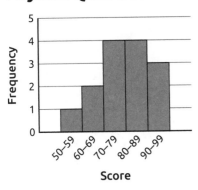

Physics Quiz Scores

1. Which of the following types of distributions is shown by this graph?

    A. normal

    B. skewed right

    C. skewed left

2. If the median of this data is 79.5, which of the following values is most likely the correct mean?

    A. 72.5

    B. 78.4

    C. 81.4

    D. 84.2

3. True or False? From this histogram, you know that there are two modes in the data—scores of 75 and 85.

    _____

Lesson 2 | Describing Distributions

Answer the questions based on the content covered in this unit. The Unit 7 Answer Key is on page 157.

Answer the questions based on the data in the table.

The 2014 RBI (runs batted in) data of ten Major League Baseball players are shown in the table.

Batter	2014 RBI
Jose Altuve	59
Michael Brantley	97
Josh Harrison	52
Jonathan Lucroy	69
Victor Martinez	103
Justin Morneau	82
Buster Posey	89
Denard Span	37
Kurt Suzuki	61
Jayson Werth	82

1. What are the mean, median, and mode of the RBI data? Round to the nearest tenth, if necessary.

   Mean: _____

   Median: _____

   Mode: _____

2. What is the range of this data?

   _____

3. Which of the following statements best describes the measures of center for this data?

   A. The mean is lower than the median because of the outlier 37.

   B. The median is higher than the mean because of the outlier 103.

   C. The mean is higher than the median because of the outlier 103.

   D. The median is lower than the mean because of the outlier 37.

Answer question 4 based on the histogram.

4. A factory manager sorts the bolts manufactured in January by diameter and graphs the results on the histogram. Given that the data set has a *perfectly normal* distribution, which statement is true?

   A. The mode is greater than the mean or median.

   B. The median is greater than the mean or mode.

   C. All three measures of center have a value of 3.0 cm.

   D. All three measures of center have a value of 3.5 cm.

Distribution of Bolts Made at Factory in January

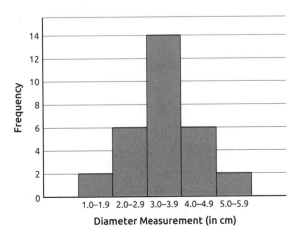

Choose the histogram that fits the following description.

5. Which distribution is skewed right because of an extremely large outlier?

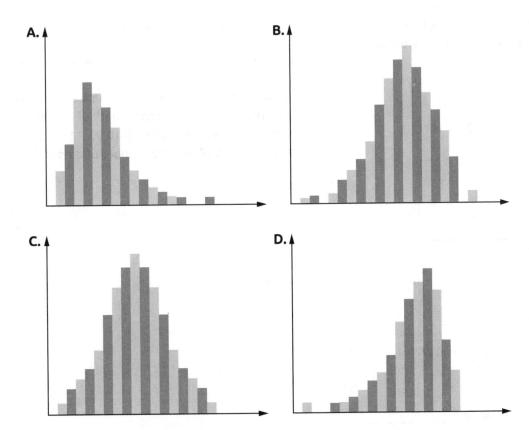

POSTTEST

Answer these questions to see how well you have learned the math content and skills in this book.

First, round each dollar amount to the nearest dollar, and then estimate the total.

1. $19.75 + $3.39 = _____

First, round each dollar amount to the nearest ten dollars and then estimate the total.

2. $6.50 + $11.39 + $25.25 = _____

First, round each dollar amount to the nearest hundred dollars and then estimate the difference.

3. $1,764 − $418 = _____

Solve the problem.

4. Rectangle ABCD is partially drawn on the coordinate grid shown. The rectangle is 5 units wide. If point C is in Quadrant I and point D is in Quadrant IV, what are the coordinates of point C and point D?

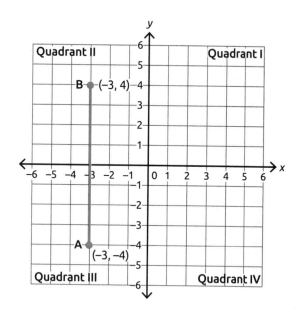

Point C: _____   Point D: _____

Answer these questions based on the data in the table.

An organization accepts donations of hair longer than 10 inches to be made into wigs for cancer patients. To help out, Tabitha got 19 of her classmates to join her in donating their ponytails to the organization. The table shows the lengths, in inches, of the girls' ponytails.

12	15	14	13.25	15.25	14.25	12.25	13	12.25	15.75
10	11	15	12	11.5	12	11.75	11.75	13.25	13.25

5. In inches, what are the mean, median, and mode of the data? Round your answers to the nearest hundredth, if necessary.

Mean: _____

Median: _____

Mode(s): _____

6. What is the range of the data?

_____

7. Based on the data in the table, create a histogram. Label the horizontal axis the length of ponytail (in inches). Use ranges like 10–10.99, 11–11.99, etc. Label the vertical axis the number of girls with a scale of 1 to 5.

8. Which of the following statements about the measures of center for the data is true?

   A. The value 15.75 inches increases the median but has no effect on the mean.

   B. The value 15.75 inches has no effect on the mode.

   C. The value 10 inches has no effect on the median.

   D. The value 10 inches has no effect on the mean.

Solve the problems.

9. Yetta had a gift card for a department store. She bought 4 scarves that each cost $19.25 each. After the gift card, the cost of the 4 scarves plus $8.75 tax came to $35.75. What was the value of the gift card?

   A. $25

   B. $50

   C. $75

   D. $100

10. A shipping crate contains 10 boxes of cookies. Each box contains 28 cookies. How many cookies are shipped if a truck can carry 85 crates?

    A. 280

    B. 2,380

    C. 23,800

    D. 230,800

Answer the question based on the illustration.

11. Look at the time on the clock shown and determine what the time was 35 minutes earlier.

    _____

Compare the two numbers using the given number line. Write < (less than), > (greater than), or = (equal to) in the blank to make each statement true.

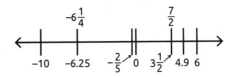

12. 0 _____ −10

13. −6.25 _____ 6

14. $3\frac{1}{2}$ _____ $\frac{7}{2}$

15. 4.9 _____ −10

Solve the problems.

16. There are 52 white keys and 36 black keys on a piano. Write the ratio of black keys to the total number of keys on a piano.

    A. 9:13

    B. 13:9

    C. 22:9

    D. 9:22

17. Scientists approximate that humans first used stone tools around 2.4 million years ago. Write the standard form of this number.

    _____

18. The furniture in a cafeteria consists of chairs and tables. There are 8 chairs for every table. If the cafeteria has a total of 54 pieces of furniture, how many pieces are chairs?

    A. 6

    B. 9

    C. 48

    D. 62

Answer the questions based on the bar graph shown.

A Chicago news program uses this graph to show rush–hour travel speed on four expressways in the area.

Rush-Hour Travel Speeds

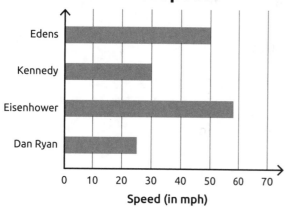

19. What is the travel speed on the slowest route?

    _____ mph

20. Which expressway has the fastest travel speed?

    _____

21. The Edens Expressway is twice as fast as which other expressway?

    _____

22. Dana is driving on the Kennedy Expressway during rush hour. Her total driving distance is 16.5 miles. How many minutes will it take her to drive the distance on the Kennedy?

    _____ minutes

Solve the problems.

23. Write "The quotient of eighteen and the sum of seven and *T*" as an expression.

    _____

24. Nora is thinking of a number. She says that if you add 9 to her number, then multiply the sum by 7, the result will be −70. What number is Nora thinking of?

    _____

25. Raphael is thinking of a number. He says that if you divide his number by 8, then subtract 5, the result will be −1. What number is Raphael thinking of?

    _____

Solve the problems.

26. Determine the temperature inside an oven in degrees Fahrenheit as indicated by the thermometer shown here.

    A. 190°F   **C.** 380°F

    B. 195°F   **D.** 385°F

27. Starting on the first day of the month, Alec works from home every sixth day, while Martin works from home every fourth day. On which days of the month do both Alec and Martin work from home?

    A. the 6th, 12th, 18th, and 24th

    B. the 12th and the 24th

    C. the 15th and 30th

    D. the 12th and 30th

28. A hardware store rents lawn mowers, snow blowers, and generators. Calculate the missing information in the inventory table, indicated by the letters A, B, and C.

Type of Machinery	Number Owned by the Store	Number Available for Rent	Percent Available for Rent
Lawn mowers	A	30	40%
Snow blowers	80	62	B
Generators	125	C	75.2%

Choose the expression that could be used to solve the problem.

29. Last week, Santiago studied three more than half as many hours for chemistry class as he did for math class. If he studied $5\frac{1}{2}$ hours for math, how many hours did Santiago study for chemistry?

    A. $5\frac{1}{2} \div 2 + 3$

    B. $3 + 5\frac{1}{2} \times 2$

    C. $5\frac{1}{2} \div \frac{1}{2} + 3$

    D. $3 + \frac{1}{2} \div 5\frac{1}{2}$

Solve the problems.

30. Convert 0.025 miles to feet.
    (Hint: 1 mile = 5,280 feet)

    _____ ft

31. Convert 17.4 centimeters to kilometers.

    _____ km

32. On the continent of Africa, the area of Tanzania is 975,300 square kilometers, while the area of Zambia is 752,610 square kilometers. Approximately how many square kilometers larger is Tanzania than Zambia? Round to the nearest hundred thousand to estimate.

    A. 100,000

    B. 150,000

    C. 200,000

    D. 250,000

Use the diagram to solve the problem.

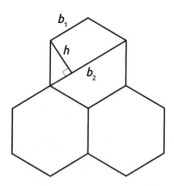

33. The honeycombs of bees are in a hexagonal pattern to maximize strength and capacity. In the illustration, notice that each hexagon can be thought of as two trapezoids with height h and bases b_1 and b_2. If the height $h = 4$ mm, $b_1 = 5$ mm, and $b_2 = 11$ mm, how many square millimeters is the area of the 3 hexagons shown? Use the formula for the area of a trapezoid: $A = \frac{1}{2} h(b_1 + b_2)$.

    A. 64

    B. 96

    C. 192

    D. 384

Solve the problem.

34. Felicia drove 3,051.5 miles from Seattle, Washington, to Boston, Massachusetts, in 4 days. How many miles did she drive per day if she drove exactly the same number of miles each day?

    _____ mi

Fill in the blanks based on the illustration. Assume that line EG is perpendicular to line AC. Use three letters to name any angles.

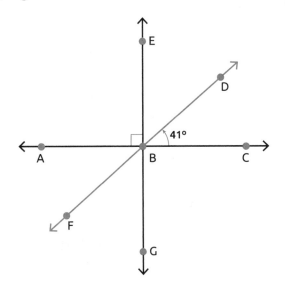

35. ∠ABE and ∠ _____ are supplementary angles.

36. ∠CBD and ∠ _____ are complementary angles.

37. m∠EBC = _____

38. m∠DBE = _____

39. m∠ABF = _____

40. Which of the following is the standard form of the number shown in expanded form?

$(3 \times 1,000) + (5 \times 10) + (8 \times 1) + (4 \times \frac{1}{10}) + (1 \times \frac{1}{1,000})$

A. 3,058.401

B. 3,580.4001

C. 30,058.41

D. 300,580.401

Answer the questions based on the data in the table.

The manager of a shoe store recorded the shoe sizes purchased by women over the last week.

5								
6	6.5							
7	7	7	7	7.5	7.5			
8	8	8	8	8	8	8	8.5	8.5
9	9	9	9.5					
10								

41. What are the mean, median, and mode of the shoe size data? Round to the nearest tenth, if necessary.

Mean: _____   Median: _____   Mode: _____

42. Based on the distribution of the shoe sizes, which of the following statements best describes the measure of center? (NOTE: You may wish to sketch a histogram for the data.)

A. The median is a better measure of center for this set of data.

B. The mean is higher than the median or the mode because of the outlier 10 and thus is not a good measure of center.

C. The median is lower than the mean because of the size 5 shoes and thus is not a good measure of center.

D. The mean, median, and mode are all accurate measures of center for this set of data.

43. Which of the following is the expanded form of 12.07?

A. $(1 \times 10) + (2 \times 1) + (7 \times \frac{1}{10})$

B. $(2 \times 10) + (1 \times 1) + (7 \times \frac{1}{100})$

C. $(1 \times 10) + (2 \times 1) + (7 \times \frac{1}{100})$

D. $(1 \times 100) + (2 \times 10) + (7 \times 1)$

Solve the problem.

44. Sally's total restaurant bill, including 10% tax and a $3 tip, came to $19.60. What percent tip did she leave the server? Round your answer to the nearest whole percent.

   A. 10%

   B. 15%

   C. 18%

   D. 20%

Evaluate the following expressions for the given values of the variable or variables.

45. $D - \dfrac{3}{A^2}$ when $D = 4$ and $A = -1$ _____

46. $-x(2.2y + 7)$ when $x = -0.8$ and $y = 4$ _____

Solve the problem.

47. Jaquez is saving money to buy a boat. He made an initial deposit of $800 and has since saved 35% of the remaining cost of the boat. If he has saved $1,400, how much does the boat cost?

   A. $4,000

   B. $4,800

   C. $5,200

   D. $5,600

Solve the problem.

48. Find the perimeter of the pentagon shown. Note: All sides of the pentagon are the same length.

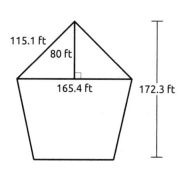

115.1 ft
80 ft
165.4 ft
172.3 ft

_____ ft

Answer the following questions based on the data in the table.

The table lists the weights, in pounds, of 25 dogs available for adoption from a local shelter.

15	99	65	79	53
35	108	88	81	67
45	65	31	90	66
80	84	17	102	94
119	32	54	62	26

49. What is the mode of the data? _____ lb

50. What is the range of the data? _____ lb

51. Which of the following statements best describes the measures of center for this data?

   A. There are no outliers, or extreme values, in this set of data.

   B. The range indicates that the weights vary greatly.

   C. A mean weight of 66 pounds shows that the mode is not a reliable measure of center.

   D. The median is the best measure of center for a data set with an odd number of values.

Solve the problems.

52. An essay contains 549 words. To the nearest hundred, how many words are in the essay?

   _____

53. A city has a population of 2,998,329. To the nearest ten thousand, how many people live in the city?

   _____

Solve the problem.

54. Three sections of pipe measuring 8.7 centimeters, 7.8 centimeters, and 9.4 centimeters are welded together to form a longer pipe. How long is the new pipe?

_____ cm

Match each fraction in column A with the equivalent decimal in column B.

Column A	Column B
55. $\frac{110}{-4}$	undefined
56. $-\frac{40}{40}$	−0.075
57. $-7\frac{1}{40}$	−0.175
58. $\frac{-7}{40}$	−1
59. $-\frac{3}{40}$	−7.025
60. $-\frac{40}{0}$	−27.5

61. Grandma Anne has coins that she wants to distribute equally to her grandchildren. She has 75 nickels, 45 dimes, and 120 quarters. To how many grandchildren can she give the coins, and how many of each coin will each grandchild receive?

_____

Solve the problem.

62. A hamburger restaurant is famous for its $\frac{3}{4}$-pound beef burgers. How many burgers can be made in one week (7 days) if the restaurant uses 108 pounds of beef each day?

A. 144

B. 567

C. 1,008

D. 3,024

Solve the problem.

63. It takes 9 bricklayers 42 hours to build a chimney. How many hours would it take for 6 bricklayers to build a chimney?

A. 28

B. 36

C. 49

D. 63

Solve each equation.

64. $20x = \frac{48}{11}$ _____

65. $2.7p + 3.3 = 8.7$ _____

66. $4n - 7 + 3 = -n + 8 - n$ _____

Solve the problem.

67. The scale on a map indicates that $1\frac{1}{2}$ inches = 40 miles. If the driving distance between Macon, Georgia, and Montgomery, Alabama, is 185 miles, about how many inches would this distance be on the map? Round your answer to the nearest inch.

_____ in

Solve each inequality.

68. $\frac{1}{4} - d \geq \frac{3}{4}$ _____

69. $\frac{x}{0.5} + 1 < 85$ _____

70. $-2y + 7 - 4 > 6y - 1 - 4y$ _____

Solve the problems.

71. Hyeon is a graphic artist who works either in a computer lab or in her office. During a 50-hour work week, Hyeon works in the computer lab for $18\frac{7}{10}$ hours. How many hours does Hyeon work in her office? Express your answer as a mixed number.

   _____ hours

72. A bowl is in the shape of a half sphere. How many cubic inches can the bowl hold if its diameter measures 18 inches? Use 3.14 for π and round your answer to the nearest whole number. Use the formula for volume of a sphere: $V = \frac{4}{3}\pi r^3$.

   A. 4,578

   B. 3,052

   C. 1,526

   D. 763

73. Lise can crochet 3 quilt squares in $\frac{2}{3}$ of an hour. How many quilt squares can she crochet in 6 hours?

   _____ quilt squares

74. Solve the inequality $-4 - \frac{2c}{3} < 0$, and then sketch its graph using the number line provided.

   _____

   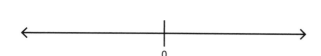

   0

75. Tom rode his motorcycle 28 miles in half an hour. Rewrite this as a unit rate of miles per hour.

   A. $\dfrac{1\,hour}{28\,miles}$ C. $\dfrac{1\,hour}{56\,miles}$

   B. $\dfrac{28\,miles}{1\,hour}$ D. $\dfrac{56\,miles}{1\,hour}$

76. Imani has at most $10 to spend on school supplies. Packs of paper are $2 each, and folders are $0.50 each. Imani has to buy 6 folders. How many packs of paper can she afford to buy with the remaining money?

   A. 4 or fewer

   B. 3 or fewer

   C. 2 or fewer

   D. at least 4

77. What is the product of 6 and k decreased by 4 when $k = \frac{2}{3}$?

   _____

78. Which of the following statements is true?

   A. A straight angle is an obtuse angle.

   B. A straight angle is an acute angle.

   C. A straight angle measures 360°.

   D. A straight angle and an acute angle are supplementary.

79. Which of the following statements is true?

   A. Parallel lines intersect at exactly one point.

   B. Perpendicular lines cross to form a right angle.

   C. Two lines in space extend forever without touching.

   D. Straight lines can intersect at exactly two points.

148

Solve the problems.

Answer the question based on the illustration.

80. The mass of Sammy, a male African elephant, is 680 kilograms less than three times the mass of Ursula, a female Asian elephant. Ursula's mass is 2,167 kilograms. What is Sammy's mass in grams?

   A. 4,461

   B. 5,821

   C. 4,461,000

   D. 5,821,000

81. The coordinate grid shows the street layout of a town, where 1 unit equals 1 block. Elvis starts at the bookstore at point B and walks 3 blocks west, 7 blocks north, then 1 block west to his apartment. Plot point A to show the location of Elvis's apartment. Label the coordinates.

82. Look at the glass beaker shown and estimate how many more milliliters of water need to be added to fill the beaker to the 250 ml line.

   A. about 15 ml

   B. about 20 ml

   C. about 35 ml

   D. about 50 ml

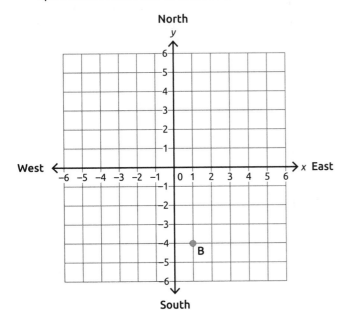

Answer the questions based on the line graph shown.

Account Balances

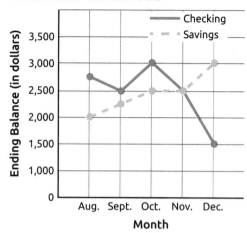

83. At the beginning of which month were the checking and savings balances equal? _____

84. What was the difference between the account balances in August? _____

85. Between which two months was there the greatest change in the checking account balance?

_____

86. Between which two months was there the least change in the savings account balance?

_____

87. What trend do you see for the savings account balance? _____

Solve the problem.

Type of Music	Tally
Rock	卌 卌 卌 卌 卌 卌 卌 卌 卌 卌
Pop	卌 卌
Rap	卌 卌 卌 卌 卌 卌 卌 卌 卌 卌 卌
Country	卌 卌 卌 卌 卌 卌 卌

88. Out of 250 students, find the percentage who prefers each type of music and create a bar graph comparing the percents.

Percent

Type of Music

POSTTEST ANSWER KEY

1. **$23**; To round dollar amounts to the nearest one, determine if the value in the tenths place is 5 or more. $19.75 rounds up to $20, and $3.39 rounds down to $3, so $20 + $3 = $23 is an estimate for the total.

2. **$50**

3. **$1,400**

4. **C: (2, 4); D: (2, −4)**; Start at point B and move 5 units to the right to find point C. Start at point A and move 5 units to the right to find point D.

5. **Mean: 12.93 in; Median: 12.63 in; Mode(s): 12 in and 13.25 in**

6. **5.75 in**; The range is the difference between the least and greatest values.

7.

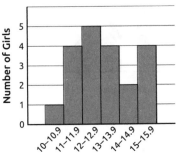

Hair Donations by Length of Ponytail

Length of Ponytail (in inches)

8. **B.** The mode is the value that appears the most. The value 15.75 inches does not affect the mode, which is 12 inches.

9. **B.**

10. **C.**

11. **8:30**; The clock shows the time 9:05, so 35 minutes earlier it was 8:30.

12. **>**

13. **<**

14. **=**; Since $3\frac{1}{2}$ and $\frac{7}{2}$ are at the same place on the number line, the values are equal.

15. **>**

16. **D.**

17. **2,400,000**

18. **C.** The ratio of chairs to total pieces of furniture is 8:9. Use the ratio to write a proportion:
$$\frac{8 \text{ chairs}}{9 \text{ total pieces}} = \frac{? \text{ chairs}}{54 \text{ total pieces}}.$$
Multiply, then divide, to find the number of chairs: $8 \times 54 \div 9 = 48$.

19. **25 mph**

20. **Eisenhower**

21. **Dan Ryan.** The Edens Expressway has a travel speed of 50 mph, which is twice the travel speed of the Dan Ryan Expressway, which is 25 mph.

22. **33 minutes**

23. $\frac{18}{7 + T}$ or $\frac{18}{T + 7}$

24. **−19** Write an equation that describes Nora's number, n: $(n + 9) \times 7 = -70$. Divide both sides of the equation by 7:
$$(n + 9) \times \frac{7}{7} = -\frac{70}{7}$$
Subtract 9 from both sides to determine n:
$$n + 9 - 9 = -10 - 9 = -19$$

25. **32**

26. **C.** The marks between 300 °F and 400 °F each represent 10 °F, so the thermometer reads 380 °F.

27. **B.**

28. **A = 75, B = 77.5%, C = 94**

29. **A.**

30. **132 ft** Use the conversion equation 1 mile = 5,280 feet. Since a mile is larger than a foot, multiply: $0.025 \times 5,280 = 132$ feet.

31. **0.000174 km**

32. **A.**

33. **C.** The 3 hexagons have the same area as 6 trapezoids, so multiply the formula for the area of a trapezoid by 6:
$$A = 6 \times \frac{1}{2} h(b_1 + b_2) = 3h(b_1 + b_2)$$
Substitute values for height and bases:
$$3h(b_1 + b_2) = 3(4)(5 + 11) = 192 \text{mm}^2$$

34. **762.875 mi**

35. **∠EBC or ∠ABG or ∠CBG**

36. **∠DBE or ∠FBG**

37. **90°**; ∠EBA is a right angle, so it measures 90°. Since ∠EBA and ∠EBC are supplementary angles, the sum of their measures must equal 180°, so m ∠EBC = 90°.

38. **49°**

39. **41°**

40. **A.**

41. **Mean: 7.8, Median: 8, Mode: 8**

42. **D.** The mean is approximately 7.8, and the median and mode are both 8. Since the mean, median, and mode are all so close, they are all accurate measures of center for this set.

43. **C.**

44. **C.**

45. **1** Substitute D = 4 and A = −1 in the equation: $D - \frac{3}{A^2} : 4 - \frac{3}{(-1)^2} = 4 - 3 = 1$.

46. **12.64**

47. **B.**

48. **575.5 ft**; Since all sides of the pentagon are the same length, multiply the length of one side by 5: $115.1 \times 5 = 575.5$ ft.

49. **65**; The mode is the value that appears the most often in a data set.

50. **104**

51. **B.**

52. **500**

53. **3,000,000**

54. **25.9 cm**; To find the total, line up the digits using place value, then add: $8.7 + 7.8 + 9.4 = 25.9$ cm.

55. **−27.5**

56. **−1**

57. **−7.025**

58. **−0.175**

59. **−0.075**

60. **undefined**

61. **15 grandchildren will each receive 5 nickels, 3 dimes, and 8 quarters.** Make lists of the factors for 75, 45, and 120 to determine the GCF, 15. Divide 75, 45, and 120 each by the GCF to find how many of each coin each grandchild receives: $75 \div 15 = 5$ nickels, $45 \div 15 = 3$ dimes, and $120 \div 15 = 8$ quarters.

62. **C.**

63. **D.** Set up a proportion that compares the bricklayers' rates: $\frac{\frac{1}{9} \text{job}}{42 \text{ hours}} = \frac{\frac{1}{6} \text{job}}{?}$. To solve, multiply 42 by $\frac{1}{6}$, then divide by $\frac{1}{9}$:
$$42 \times \frac{1}{6} \div \frac{1}{9} = \frac{42}{6} \times \frac{9}{1} = 63.$$

64. $x = \frac{12}{55}$

65. **p = 2**; First, subtract 3.3 from both sides of the equation:
$$2.7p + 3.3 - 3.3 = 8.7 - 3.3$$
Then divide both sides by 2.7:
$$\frac{2.7p}{2.7} = \frac{5.4}{2.7} = 2$$

66. **n = 2**

67. **7 inches**; Set up a proportion comparing map length to actual distance: $\frac{\frac{11}{2} \text{inches}}{40 \text{miles}} = \frac{? \text{ inches}}{185 \text{ miles}}$.
Multiply, then divide to solve:
$$\frac{11}{2} \times 185 = \frac{3}{2} \times \frac{185}{1} = \frac{555}{2}; \ \frac{555}{2} \div 40$$
$$= \frac{555}{2} \times \frac{1}{40} = \frac{555}{80} \approx 7 \text{ inches}$$

68. $d \leq -\frac{1}{2}$ or $-\frac{1}{2} \geq d$

69. **x < 42 or 42 > x**; Subtract 1 from both sides of the inequality:
$$\frac{x}{0.5} + 1 - 1 < 85 - 1.$$ Multiply both sides of the inequality by 0.5: $\frac{x}{0.5}$ $0.5 < 84$ 0.5; $x < 42$.

70. $y < 1$ or $1 > y$

71. $31\frac{3}{10}$ hours

72. **C.** Since the bowl is in the shape of half a sphere, multiply the formula for the volume of a sphere by $\frac{1}{2}$.

$V = \frac{1}{2} \times \frac{4}{3}\pi r^3 = \frac{2}{3}\pi r^3$

Substitute 3.14 for π and 9 for r.

$\frac{2}{3}\pi r^3 = \frac{2}{3}(3.14)(9)^3 \approx 1{,}526\ \text{in}^3$

73. **27 quilt squares** Set up a proportion relating quilt squares and time:

$\dfrac{3\ \text{quilt squares}}{\frac{2}{3}\ \text{hour}} = \dfrac{?\ \text{quilt squares}}{6\ \text{hours}}$. to solve, multiply by 3 and 6, then divide by $\frac{2}{3}$: $3 \times 6 = 18; 18 \div \frac{2}{3} = \frac{18}{1} \times \frac{3}{2} = \frac{54}{2} = 2$.

74. $c > -6$ or $-6 < c$

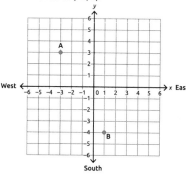

75. **D.** A unit rate always compares a quantity to 1, so write a proportion that compares a quality to 1:

$\dfrac{28\ \text{miles}}{\frac{1}{2}\ \text{hour}} = \dfrac{?\ \text{miles}}{1\ \text{hour}}$. Multiply then divide to solve: $28 \times 1 = 28$;

$28 \div \frac{1}{2} = \frac{28}{1} \times \frac{2}{1} = 56\ \text{miles}$.

76. **B.** Write and simplify an expression describing the amount of money Imani has after buying 6 folders: $10 - 6(\$0.50) \div \$7$. Then, divide the remaining \$7 by the cost of one pack of paper: $\$7 \div \$2 = 3.5$. Since packs of paper cannot be split, Imani can buy 3 or less packs of paper.

77. 0

78. **A.** A straight angle measures 180°. Since an obtuse angle measures greater than 90°, a straight angle is an obtuse angle.

79. **B.**

80. **D.** Write and simplify an expression for Sammy's weight in kilograms: $(3 \times 2{,}167) - 680 = 5{,}281$ kilograms. To change kilograms into grams, move the decimal three places to the right: 5,281 kilograms = 5,281,000 grams.

81. Point A is at $(-3, 3)$.

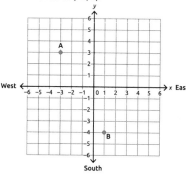

82. **C.**

83. November

84. **$750** In August, the checking balance was \$2,750 and the savings balance was \$2,000. To find the difference, subtract: $\$2{,}750 - \$2{,}000 = \$750$.

85. **November and December**

86. **October and November;** Since the savings account values stayed the same, there was no change to the savings account balance between October and November.

87. The savings account balance slowly increases over this time period.

88. Sample graph:

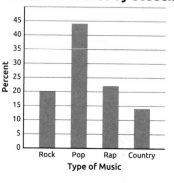

Music Preference by Students

After checking your *Posttest* answers using the *Answer Key*, use the chart below to find the questions you did not answer correctly. Then locate the pages in this book where you can review the content needed to answer those questions correctly.

Question	Where to Look for Help		
	Unit	Lesson	Page
1, 2, 3, 32, 52, 53	1	3	19
4, 81	5	2	89
5, 6, 8, 41, 42, 49, 50, 51	7	1	130
7	7	2	134
9, 10, 34, 61, 76, 80	3	1	54
11	6	2	115
12, 13, 14, 15, 26	1	1	12
16, 18, 22, 63, 67, 73	2	1	44
17, 30, 40, 43	1	2	15
19, 20, 21, 83, 84, 85, 86, 87, 88	6	4	123
23, 29	4	1	60
24, 25, 45, 46, 65, 66, 77	4	2	64
27	1	5	27
28, 44, 47	2	2	48
31	6	3	119
33, 48	5	4	99
35, 36, 37, 38, 39, 78	5	3	94
54	1	4	23
55, 56, 57, 58, 59, 60	1	6	31
62, 64, 71, 75	1	7	35
68, 69, 70	4	3	68
72	5	5	104
74	4	4	72
79	5	1	84
82	6	1	111

Answer Key

ANSWER KEY

Unit 1: Numbers and Operations

Lesson 1: Representing Numbers on a Number Line
1. <
2. <
3. >
4. <
5. −7
6. 2
7. 8
8. C.

Lesson 2: Understanding Place Value
1. C.
2. B.
3. D.
4. 40,725.8

Lesson 3: Rounding Numbers and Estimating
1. 290
2. 3,400
3. 10,000
4. 2,000
5. 84.3
6. 0.15
7. 9.527
8. 3,003
9. $14
10. $8
11. $90
12. $170
13. B.
14. B.
15. D.

Lesson 4: Performing Operations on Whole Numbers and Decimals
1. C.
2. C.
3. D.
4. 1,242

Lesson 5: Finding Common Factors and Multiples
1. B.
2. LCM: 45; GCF: 3
3. A.
4. 48

Lesson 6: Understanding Fractions
1. >
2. =
3. <
4. <
5. >
6. $\frac{1}{100} = 0.01$

$2\frac{4}{5} = 2.8$

$\frac{11}{4} = 2.75$

$\frac{6}{6} = 1$

$\frac{0}{19} = 0$

$\frac{231}{1000} = 0.231$

$2\frac{11}{50} = 2.22$

Lesson 7: Performing Operations on Fractions and Mixed Numbers
1. C.
2. B.
3. $1\frac{3}{5}$
4. 16

Unit Practice Test
1. −3
2. >; Since −9 is to the right of −14 on the number line, −9 > −14.
3. B.
4. 640,029.31
5. 760; To round to the nearest ten, look at the value in the ones place, 6. Since 6 > 5, round 756.24 up to 760.
6. 0.93
7. 34.75; First, line up all the numbers using place value. Then, add to find the total.
8. C. Divide 351 by 18 to find the least number of cabins needed. Since the quotient is 19.5, the camp director needs at least 20 cabins.
9. B.
10. 3,750
11. D.
12. 13 boxes can be filled, each with 7 screws, 3 nuts, 3 bolts, and 11 nails. To find the number of boxes, determine the GCF of 91, 39, and 143, which is 13. Divide 91, 39, and 143 each by 13 to find the number of each item in every box.
13. =
14. <
15. >
16. <
17. $-\frac{13}{10} = -1.3$
18. $-13\frac{1}{8} = -13.125$
19. $-\frac{7}{8} = -0.875$
20. −1.625; To write $\frac{-13}{8}$ as a decimal, divide 13 by 8 using long division.
21. B. To add the mixed numbers, first rewrite the fractions with the common denominator of 8. Then, add the fractions: $\frac{4}{8} + \frac{7}{8} + \frac{2}{8} = \frac{13}{8}$. Rewrite the mixed number as a fraction: $\frac{13}{8} = 1\frac{5}{8}$. Then, add the whole numbers: $1\frac{5}{8} + 2 + 3 + 4 = 10\frac{5}{8}$.
22. $7\frac{5}{8}$
23. 434 To calculate $1,206 \div \frac{3}{4}$, rewrite the problem as a multiplication problem: $\frac{1,206}{1} \times \frac{4}{3} = \frac{4,824}{3}$. Divide 4,824 by three to find the number of sections that can be made.
24. C.

Unit 2: Ratios and Proportional Relationships

Lesson 1: Using Ratios and Proportions to Solve Problems
1. 18
2. 81.25 or $81\frac{1}{4}$
3. A.
4. C.

Lesson 2: Solving Percent Problems
1. C.
2. $20.00
3. B.
4. S. Gostkowski with a percentage of 94.6%

Unit Practice Test
1. C. Use the ratio 7:1 to set up a proportion: $\frac{7 \text{ campers}}{1 \text{ counselor}} = \frac{? \text{ campers}}{37 \text{ counselors}}$. Multiply 7 and 37 to find the cross product, then divide by 1 to find the maximum number of campers.
2. 36; Since 1 minute, 30 seconds = 90 seconds, use the ratio 5:90 to set up a proportion: $\frac{5 \text{ waffles}}{90 \text{ seconds}} = \frac{2 \text{ waffles}}{? \text{ seconds}}$. Multiply 90 by 2, then divide by 5: $90 \times 2 = 180$; $180 \div 5 = 36$.
3. D.
4. No, because the cross products are not equal. $8.2 \times 0.45 \neq 2.99 \times 0.8$ or $3.69 \neq 2.392$
5. C.
6. D. Use percent × whole = part to write an equation: $85.5\% \times 200 = \text{part}$. Convert the percent to a decimal, then multiply by 200: $0.855 \times 200 = 171$.
7. A.

8. $25 8. $25; Use part ÷ percent = whole to write an equation: $4.50 ÷ 18% = whole. Convert the percent to a decimal, then divide: 4.50 ÷ 18% = 25.

9. D.

10. A.

Unit 3: Operations and Algebraic Thinking

Lesson 1: Interpreting and Solving Word Problems

1. A.
2. C.
3. A.
4. C.

Unit Practice Test

1. C.
2. B.
3. D.
4. A. Write an equation for the total ounces of cheese packed in one day: $(8 + 9 + 10) \times 6 \times 8 =$ total. Add, then multiply to find the total ounces: $(8 + 9 + 10) \times 6 \times 8 = 1,296$.
5. B.
6. B. Write an equation that describes Janelle's age: Terrie's age + Janelle's age = 24. Replace Terrie's age with $(3 \times$ Janelle's age): $(3 \times$ Janelle's age) + Janelle's age = 24, or 4 × Janelle's age = 24. Divide to find Janelle's age.

Unit 4: Algebraic Expressions, Equations, and Inequalities

Lesson 1: Evaluating Expressions

1. $7x$
2. $b + 12$
3. $-5J$
4. $g \div h$
5. $C^3 - 9$
6. $6(z - 22)$
7. $445(x + y)$
8. $-R - 74$
9. C.
10. B.
11. B.

Lesson 2: Solving Equations

1. 36
2. −15
3. −11
4. 1.4
5. −7
6. −8
7. 1
8. 8

Lesson 3: Solving Inequalities

1. $d \le 10$
2. $z < -14$
3. $y \ge -15$
4. $t < 4\frac{2}{3}$
5. $h < 0.4$
6. $v < -24$
7. $m \le -2$
8. $x \le -1$

Lesson 4: Graphing Equations and Inequalities

1.

2.

3.

4.

5.
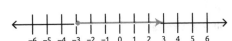

6.
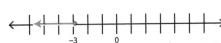

7.

Lesson 5: Solving Problems Using Expressions, Equations, and Inequalities

1. D.
2. B.
3. A.
4. B.

Unit Practice Test

1. $y + 8$
2. $h^4 - 6$
3. $24.5 - 7b$; Since "decreased by" refers to subtraction, rewrite the phrase as $24.5 - 7b$.
4. $\frac{x}{2y + z}$
5. fg^2h
6. C. Substitute $a = 12$ and $b = 10$ into the expression $a - 0.4b$; $12 - 0.4(10)$. Multiply, then subtract, to evaluate the expression: $12 - 0.4(10) = 8$.
7. D. Rewrite the phrase "eight fewer than the product of r and s" as an expression: $rs - 8$. Substitute $r = 5$ and $s = 7$: $(5)(7) - 8$. Multiply, then subtract, to evaluate the expression: $(5)(7) - 8 = 27$
8. 6
9. −2
10. $9\frac{1}{2}$
11. 4; To solve $8b - 1 = 31$, first add 1 to both sides: $8b - 1 + 1 = 31 + 1$. Then divide both sides by 8. $\frac{8b}{8} = \frac{32}{8} = 4$
12. 174
13. 23; To solve $9 - 5t = 3t - t - 14$, first group like terms:
$9 - 5t = -4t - 14$
Add $5t$ to both sides:
$9 - 5t + 5t = -4t - 14 + 5t$
Add 14 to both sides:
$9 + 14 = t - 14 + 14$
$t = 23$
14. −3
15. $s > 4.5$
16. $y \ge 22$; To solve $y - 3 \ge 19$, add 3 to both sides: $y - 3 + 3 \ge 19 + 3$; $y \ge 22$.
17. $b > -1$; To solve $85 - 10b < 95$, subtract 85 from both sides:
$85 - 10b - 85 < 95 - 85$
Then divide both sides by −10:
$\frac{-10b}{-10} < \frac{10}{-10}$
Since both sides were divided by a negative number, reverse the order of the inequality symbol: $b > -1$
18. $p \ge 16$
19. $c < -\frac{13}{4}$ or $-3\frac{1}{4}$
20. $r < 1$

ANSWER KEY

21. $a > 2$;

22. $x = 15$;

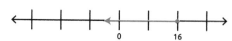

23. $c \leq 16$;

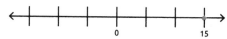

24. $y > -\frac{1}{2}$;

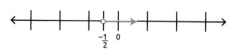

25. **C.** Write an equation that models the amount the water cooler can hold, the amount Melinda removed, and the amount that remained in the water cooler: $18.9 - 5w = 9.8$

26. **B.**

27. **C.** Write an equation that models Charlie's number, n: $\left(\frac{n}{-4}\right) - 3 = -15$
To solve, first add 3 to both sides: $\left(\frac{n}{-4}\right) - 3 + 3 = -15 + 3$
Then multiply both sides by -4:
$\left(\frac{n}{-4}\right) \times -4 = -12 \times 4$; $n = 48$

28. **A.** Write an inequality that models the amount that each student pays, c: $c \geq \frac{\$22,080}{32}$. Divide to simplify the inequality: $c \geq \$690$.

Unit 5: Geometry

Lesson 1: Knowing Shapes and Their Attributes

1. A.
2. B.
3. D.
4. C.

Lesson 2: Using a Coordinate Plane

1. (0, 3)
2. (−3, 1)
3. (−5, −5)
4. (4.5, 0) or $(4\frac{1}{2}, 0)$
5. (1.5, −2.5), or $(1\frac{1}{2}, -2\frac{1}{2})$

6–10.

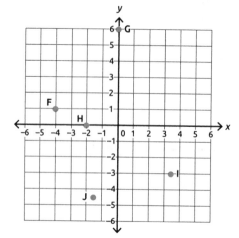

11. (175, 80)
12. C.

Lesson 3: Solving Angle Measure Problems

1. r
2. p or r
3. q or s
4. q
5. 62°
6. 14°
7. C.

Lesson 4: Solving Perimeter and Area Problems

1. 42.25 or $42\frac{1}{4}$
2. D.
3. 225
4. B.

Lesson 5: Solving Surface Area and Volume Problems

1. 220
2. C.
3. B.
4. A.

Unit Practice Test

1. **B.**
2. **B.** Rhombuses and squares are both shapes that have 4 equal sides.
3. (−35, −30)
4. (50, −20)
5. **(55, 45);** To reach Point O from the origin, first move 55 units to the right, then move 45 units up.

6. (0, −40)

7–10.

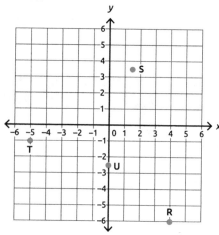

11. **(1, −4.5) or $(1, -4\frac{1}{2})$;** To find the coordinates of the house, start at R (−4, 3). Then move 5 units to the left, and 7.5 units down.

12. **false;** ∠1 and ∠3 are vertical angles, so they have the same measurement. Since none of the angles measure 90°, m ∠1 + m ∠3 ≠ 180.

13. **true;** ∠2 and ∠3 are supplementary angles, so their measures will add up to 180°. If m ∠3 = 48°, then m∠2 = (180−48)° = 132°

14. **true**

15. **false**

16. **A.**

17. **D.** Write an equation for the perimeter of a rectangle: $l + l + w + w = 24$.
Substitute $l - 8 = w$:
$l + l + (l - 8) + (l - 8) = 24$
Group like terms: $4l - 16 = 24$
Add 16 to both sides:
$4l - 16 + 16 = 24 + 16$
Divide both sides by 4: $4\frac{l}{4} = \frac{10}{4} = 10$

18. **A.** To find the area of the rectangle, use the equation $A = lw$: $A = (4)(6) = 24 \text{ cm}^2$
To find the area of the shaded triangle, use the equation $A = \frac{1}{2}bh = \frac{1}{2}(6)(4) = 12 \text{cm}^3$

19. **C.**

20. **B.** To find the surface area of a cube, use the equation $SA = 6s^2 = 6(5.7)^2 = 194.94 \text{ cm}^2$. To find 96% of the surface area, multiply $194.94 \times 0.96 \approx 187.1 \text{ cm}^2$.

21. **B.**

156

Unit 6: Measurement and Data

Lesson 1: Measuring and Estimating Length in Standard Units

1. $4\frac{7}{8}$ in
2. 12.3 cm
3. 2.9 cm
4. 29 mm

Lesson 2: Solving Problems Involving Measurement

1. 1875 ml, 1.875 l
2. C.
3. A.

Lesson 3: Converting Measurement Units

1. 9.43 decimeters
2. D.
3. B.
4. A.

Lesson 4: Representing and Interpreting Data

1. 10%
2. 144
3. 3:00 pm
4. A.

Unit Practice Test

1. $5\frac{3}{4}$ in, 14.6 cm; Since the smallest tick marks on the inches ruler each represents $\frac{1}{8}$ in, the leaf measures $5\frac{6}{8} = 5\frac{3}{4}$ inches. Since the smallest tick marks on the centimeter ruler each represents $\frac{1}{10}$ centimeter, the leaf measures $14\frac{6}{10}$ centimeters.

2. 4:07
3. C. Since 2 liters = 2,000mL, find $2,000 \div 250$ to determine the number of containers needed: $2,000 \div 250 = 8$.
4. 4.18 kg; To find the weight in grams, multiply: $19 \times 220 = 4,180$ grams. To convert 4,180 grams to kilograms, move the decimal point 3 places to the left: 4.180 kilograms.
5. 4.02678 km
6. C.
7. 10; To find the fewest number of hits, determine the shortest bar. The shortest bar is for game 2, which shows that 10 was the fewest number of hits in a game.
8. games 4 and 6
9. 1 meter
10. The plant grew quickly from April to June and then grew more slowly from July to October.

Unit 7: Basic Statistics

Lesson 1: Understanding Statistical Variability

1. Mean: 86.1°F
   Median: 83°F
   Mode: 83°F
2. C.
3. C.

Lesson 2: Describing Distributions

1. C.
2. B.
3. false

Unit Practice Test

1. Mean: 73.1
   Median: 75.5
   Mode: 82; To find the mean, add all the data values and divide by 10. To find the median, arrange the 10 data values in order, then find the mean of the two middle values. The mode is the most common value.
2. 66
3. A.
4. D.
5. A. A distribution that is skewed right will be more spread out on the right side of the graph.

GLOSSARY

12-hour clock — clock that uses the hours 1–12 and the minutes in between to track time

acute angle — an angle that measures less than 90°

adjacent angles — angles that share a common side and a common vertex

algebraic expression — a combination of math symbols, including operators, grouping symbols, numbers, and variables

angle — formed by two rays

area — amount of space inside a 2-dimensional figure

average — usually refers to the mean of a set of data

bar graph — a graph that compares data using bars of varying lengths

base — the number raised to a power; a is the base in the expression a^b

base ten — counting system that uses ten digits (0, 1, 2, 3, 4, 5, 6, 7, 8, and 9) to form multi-digit numbers

bell curve — another name for normal distribution because of its shape

circumference — the distance around a circle

coefficient — the number in front of the variable; the coefficient of $4x$ is 4

common multiple — a number into which each of the numbers in a set can be divided evenly

complementary angles — two angles whose measures add up to 90°

congruent — identical in size and shape

continuous — data that can have any value within an interval of values; continuous data is used with histograms

coordinate grid—another term for coordinate plane

coordinate plane—a plane specially designed to show the locations of points, lines, and other geometric objects.

cross product — the result of multiplying the numerator of one fraction by the denominator of another fraction

cubic units — units used to measure volume

curve — a rounded mark

data — measurements, facts, or other pieces of information that have been collected in some way

decimal — a fractional part of a whole written using a decimal point

decimal point — the mark that separates the whole number part from the decimal part in a decimal number

degree — the amount of counterclockwise rotation of one side of an angle pivoting around its vertex

denominator — the bottom number in a fraction that represents the whole amount

diameter — the distance across a circle at its widest; passes through the center of the circle; twice the circle's radius

distributed — ways that data are spread out when graphed as a histogram

dividend — the number that is being divided

divisor — the number you are dividing by

empty set symbol — $\varnothing$; used to indicate that a set contains no elements

equation — a statement, using an equal sign, showing that quantities have the same value

equivalent fractions — fractions that have the same value

estimate — to make an approximate calculation

evaluate — to find the value

expanded form — the value of a number written out using the place value of each digit; in expanded form, 749 is written as $(7 \times 100) + (4 \times 10) + (9 \times 1)$

exponent — the power to which a number is being raised; b is the exponent in the expression a^b

exponential expression — an expression that has a number or variable raised to a power; a^b, x^2, 2^3, and $(y+2)^2$ are all exponential expressions

factor — a number that divides evenly into another number; part of a multiplication fact; 1, 2, 4, 8, and 16 are all factors of 16

fraction — a part of a whole

fraction bar — the bar that separates the numerator of a fraction from the denominator; also thought of as a division symbol

gram (g) — metric unit of capacity of the mass of a solid; 1 g = 0.001 kg

graphing — a way to visually show the solution of an equation or an inequality

greatest common factor (GCF) — the largest factor that two or more numbers have in common

histogram — a type of bar graph that compares intervals of values

hour — unit of time; 1 hr = 60 min

improper fraction — a fraction in which the numerator is equal to or larger than the denominator

inequality — the statement that two values or expressions are not equal

integers — the set of numbers that includes positive numbers, negative numbers, and zero: {... −3, −2, −1, 0, 1, 2, 3, ...}

intersection — the point at which two lines cross

inverse operations — operations that undo each other

kilogram (kg) — metric unit of capacity of the mass of a solid; 1 kg = 1,000 grams

least common multiple (LCM)—the smallest multiple that tow or more numbers have in common.

like fractions — fractions with the same denominator

like terms — terms that have the same variable(s) raised to the same power; $6x^2$ and x^2 are like terms; $20xy$ and $7xy$ are like terms

line — a straight mark that extends in both directions indefinitely

line graph — a graph that connects data points with a line to show trends over time

line segment — a line that has two endpoints

liter (l) — metric unit of measure of a gas or a liquid; 1 l = 1,000 ml

lowest terms — a fraction in which the numerator and denominator do not share a common factor

mass — amount of matter that makes up an object

mean — the sum of the values in a set of data divided by the number of values in the set

measure of center—the value that represents the middle of a set of data

median — the middle number in an ordered set of data

metric system — measurement system used throughout the world; a base ten system that includes the units centimeters, meters, grams, kilograms, milliliters, and liters

milligram (mg) — metric unit of capacity of the mass of a solid; 1 mg = 0.001 g

milliliter (ml) — metric unit of measure of a gas or a liquid; 1 ml = 0.001 l

minute — unit of time; 1 min = 60 sec

mixed number — a number that has a whole number part and a fractional part

mode — the value that appears most often in a set of data

multiple — a number that is evenly divisible by another number; 32 is a multiple of 16

natural numbers — the numbers used for counting things: {1, 2, 3, ...}; also known as the set of positive integers

negative slope — a line that falls from left to right on a graph or chart

normal distribution — where the mean, median, and mode are all equal, and located in the center of the distribution, and symmetrical.

number line — a tool for viewing and comparing numbers

numerator — the top number in a fraction that represents the part of the whole

obtuse angle — an angle that measures greater than 90°

order of operations — rules that describe the sequence in which operations (multiplication, division, etc.) should be done

ordered pair—(x, y)

origin—intersection of the x and y axes

outlier — an extreme value in a set of data

parallel lines — lines that remain equidistant from each other and never cross

percent equation — percent × whole = part

perimeter — the total distance around a straight-sided figure

perpendicular lines — lines that intersect at a 90° angle

place value — the value of a digit, depending on its place in a number

place value chart — a chart that shows the place value of each digit in a base ten number

plane — a flat, 2–dimensional surface that extends in all directions indefinitely

point — a location in space

GLOSSARY

positive integers — the set of numbers greater than 0

positive slope — a line that rises from left to right has a positive slope

prime number — a number greater than 1 that is divisible only by itself and 1

proportion — an equation that shows two ratios are equal

quadrants—the four parts that make up a coordinate plane

quotient — the answer to a division problem

radius — the distance from a point on a circle's circumference to the center of the circle; half the circle's diameter

range — the difference between the least and greatest values in a set of data

rate — a special type of ratio that compares two different units

ratio — a way to compare two quantities

ray — a line that extends in only one direction indefinitely

reciprocal — switching the numerator and denominator of a fraction; a fraction multiplied by its reciprocal equals 1

reduce — to simplify a fraction

regular hexagon — 6–sided figure with all sides equal and all angles equal

regular pentagon — 5–sided figure with all sides equal and all angles equal

right angle — a 90° angle

right triangle — triangle with one 90° angle

round — to estimate an amount to a certain place value

ruler — a straight object marked at regular intervals that is used to measure length or distance

set — a group of numbers

set-builder notation — a way of describing what is in a set; $\{x \mid x < 5\}$ is "the set of all x such that x is less than 5"

sides—line segments that make up a shape

solution set — all the numbers that make an equation or inequality true

solve — to find the solution

square units — units used to measure area

standard form — a number written in numerical form; in standard form, seven hundred forty-nine is 749

statistics — the science of collecting data in order to analyze, interpret, and display it

straight angle — a 180° angle; a straight line

supplementary angles — two angles whose measures add up to 180°

surface area — total space needed to cover a 3-dimensional object

theorem — a proven principle

undefined — describes a fraction that has a denominator of 0

unit rate — a ratio that always compares a quantity to 1; for example, miles per hour (mph) is a unit rate comparing the number of miles traveled to 1 hour of time spent traveling

U.S. customary system — measurement system used primarily in the United States; units of measure include inches, feet, ounces, pounds, fluid ounces, and gallons

variable — usually a letter of the alphabet that is used to represent an unknown value in an algebraic expression or equation

vertex — the endpoint where two sides meet

vertical angles — pairs of angles across from one another that are formed by intersecting lines; vertical angles are equal

volume — amount of space inside a 3-dimensional object

whole numbers — the natural numbers along with 0: {0, 1, 2, 3, …}

x-axis—the horizontal axis of a coordinate plane

x-coordinate—first number in the ordered pair, tells how far right or left to travel along the x-axis, starting from the origin.

y-axis—the vertical axis of a coordinate plane

y-coordinate—second number in the ordered pair, tells how far to move up or down along the y-axis, starting from the origin.

zero slope — a line that is horizontal (flat) has a slope of 0